Slo

T0252164

Slow Cures and Bad Philosophers

Essays on Wittgenstein, Medicine, and Bioethics

Edited by Carl Elliott

Duke University Press   Durham and London 2001

© 2001 Duke University Press
All rights reserved
Printed in the United States of America on acid-free paper ∞
Typeset in Stone Serif by Wilsted & Taylor Publishing Services
Library of Congress Cataloging-in-Publication Data appear
on the last printed page of this book.

In philosophizing we may not *terminate* a disease of thought. It must run its natural course, and *slow* cure is all important. (That is why mathematicians are such bad philosophers.)—Ludwig Wittgenstein, *Zettel*

Contents

Acknowledgments. I am very grateful for the excellent research assistance of Robert Crouch and Dale Turner in preparing this book, which was supported by a grant from Fonds pour la Formation de Chercheurs et l'Aide à la Recherche (Quebec).

Abbreviations

The following conventional abbreviations have been used to refer to Wittgenstein's works in the text and notes.

BB   *Preliminary Studies for the "Philosophical Investigations" Generally Known as "The Blue and Brown Books."* Oxford: Basil Blackwell, 1958.

CV   *Culture and Value.* Translated by Peter Winch. Edited by G. H. von Wright. Chicago: University of Chicago Press, 1980.

LE   "Wittgenstein's Lecture on Ethics." *Philosophical Review* 74 (Jan. 1965): 3–12.

LFM   *Wittgenstein's Lectures on the Foundations of Mathematics, Cambridge, 1939.* Ithaca, N.Y.: Cornell University Press, 1976.

NB   *Notebooks, 1914–1916.* Translated by G. E. M. Anscombe. Edited by G. H. von Wright and G. E. M. Anscombe. Chicago: University of Chicago Press, 1979.

OC   *On Certainty.* Translated by Denis Paul and G. E. M. Anscombe. Oxford: Basil Blackwell, 1969; New York: Harper and Row, 1972.

PI   *Philosophical Investigations.* Translated by G. E. M. Anscombe. Oxford: Basil Blackwell, 1953, 1958, 1968.

PO   *Ludwig Wittgenstein: Philosophical Occasions, 1912–1951.* Edited by James Carl Klagge and Alfred Nordmann. Indianapolis: Hackett, 1993.

RFM   *Remarks on the Foundations of Mathematics.* Translated by G. E. M. Anscombe. Oxford: Basil Blackwell, 1956.

RPP   *Remarks on the Philosophy of Psychology.* 2 vols. Translated by G. E. M. Anscombe. Edited by G. E. M. Anscombe and G. H. von Wright. Chicago: University of Chicago Press, 1980.

T   *Tractatus logico-philosophicus*. Introduction by Bertrand Russell. London and New York: Routledge, 1992.

WWK *Ludwig Wittgenstein und der Wiener Kreis*. By Friedrich Waismann. Frankfurt am Main: Suhrkamp, 1967.

Z   *Zettel*. Edited by G. E. M. Anscombe and G. H. von Wright. Oxford: Basil Blackwell, 1967, 1981.

1 Introduction: Treating Bioethics
Carl Elliott

What would Ludwig Wittgenstein have made of professional bioethics? It is not hard to guess. Wittgenstein loathed professional philosophy. He called it a "a kind of living death." [1] He referred to Oxford as the influenza zone and once compared a meeting of philosophical societies at Cambridge to an outbreak of bubonic plague.[2] When his student Maurice Drury was rejected for a lectureship in philosophy in favor of Dorothy Emmett and instead went to work in a Welsh garden cooperative for unemployed miners, Wittgenstein told him that he owed a great debt to Emmett, because she had saved him from becoming a professional philosopher." [3] Part of Wittgenstein's attitude toward professional philosophy was rooted in his disdain for academic life. Universities not only smothered creative thought, Wittgenstein believed, they made it difficult to be a decent human being. He claimed to prefer the gossip of his college bedmaker to the artificial conversation at the Cambridge high table. But he was only marginally less contemptuous of philosophy written for a popular audience. According to Wittgenstein, the only thing worse than professional philosophy was journalism.[4] Bertrand Russell's books of popular philosophy, for example, he despised. *The Conquest of Happiness* was a "vomative." [5]

Not exactly grounds for optimism about a book like this one. An uphill battle lies ahead for any book that claims to find useful guidance from Wittgenstein for bioethics and the practice of medicine. His personal feelings about the academic life aside, Wittgenstein simply produced very little work that dealt directly with ethics. Bioethics as an academic discipline did not get its start until years after his death. The pairing of Wittgenstein with bioethics will seem even more incongruous to those who know anything of his life and character. The media-savvy, money-driven world of contemporary U.S. bioethics could hardly be further from the world Wittgenstein inhabited, nor could it be much further away from his austere ethical sensibility. For Wittgenstein, ethics was an intensely personal, deeply serious affair. It was not even simply about good conduct and good character, but about the sense of life, the state of one's soul, or as he often put it, about being decent. The form that bioethics often takes—as a kind of anonymous, impersonal, rule-writing exercise in which we advise others how to behave—could hardly be more alien to this sort of interior ethical quest.

Certainly Wittgenstein and bioethics make a very odd pair. Yet the match may not be quite as eccentric as it initially seems. While it is true, for example, that Wittgenstein produced little formal writing on ethics, there is also a powerful sense in which ethics permeates the entirety of his work. Indeed, he claimed that the point of the *Tractatus logico-philosophicus*, to outward appearances a dense treatise on logic and language, was fundamentally ethical. And while in the *Tractatus* Wittgenstein relegated ethics to the realm of the unsayable, later in his life he rejected his Tractarian views on language. Some of his later work, such as the "Lecture on the Freedom of the Will," points toward the direction his later philosophical writing on ethics may have taken had he lived longer.

Wittgenstein also worked in a hospital during World War II, and at one time seriously considered giving up his philosophy post to go to medical school.[6] This might seem like a relatively trivial fact about Wittgenstein's life, but it is connected to a deeper part of his sensibility, and that is his desire to do work that was useful.[7] He discouraged his students from going into philosophy and sometimes urged them to do manual work instead.[8] He himself gave up philosophy after writing the *Tractatus* and became a schoolteacher to the children of peasants in rural Austria. This need to do useful work seemed to extend to his vision of philosophy as well. In a letter to Norman Malcolm, Wittgenstein wrote, "[W]hat is the use of studying philosophy if all that it does for you is to enable you to talk with some plausibility about some abstruse questions of logic, etc., & if it does not improve your thinking about the important questions of everyday life."[9]

Bioethics, even more than most areas of philosophy, aims to be useful. Even the most speculative work in bioethics is intended to improve our thinking about everyday life. Thus, the real test of a book such as this one is not Wittgenstein's personal views or his character, or even what he might have thought about bioethics, but how useful the book is. Can Wittgenstein's work help us think better and more clearly about bioethics and medical practice? If so, how?

There is probably no single uncontroversial answer to these questions, but one possibility needs to be mentioned first so that it can be put to rest. The point of this book is not to determine what Wittgenstein's opinion would have been on various problems in bioethics. That kind of book would run the danger of becoming a quasi-theological dispute ("What would the great man have thought about legalizing euthanasia?") and even at its best would have become just another collection of scholarly criticism on Wittgenstein's philosophy. The point of this book is not to examine Wittgenstein's personal moral convictions or even Wittgenstein the historical figure, but rather to explore the question of whether a deep engagement with his work can illuminate some of the problems that medicine and biological science place in front of us. And while it would be misleading to suggest that the essays collected in this book offer a unified answer to that question (some of the essayists are in striking disagreement with one another), several themes and ideas gesture toward the ways in which Wittgenstein can help us think about bioethics.

Theory and Anti-Theory

Perhaps the most thoroughgoing of these themes relates to Wittgenstein's later views on philosophy itself: his hostility toward philosophical "theory" and, in particular, toward a quasi-scientific model of philosophy. In his later work Wittgenstein rejected the notion that the philosopher's task is to build explanatory philosophical systems. In the *Philosophical Investigations*, he writes, "And we may not advance any kind of theory. There must not be anything hypothetical in our considerations. We must do away with all explanation, and description alone must take its place" (PI § 109). As for philosophy, so too for moral philosophy: "If I needed a theory in order to explain to another the essence of the ethical, the ethical would have no value at all" (WWK, p. 116–17). The danger here is that of seeing philosophy as an impersonal, objective science, in which philosophers gather data and construct theories that can then be analyzed, argued about, tinkered with, and revised.[10] For Wittgenstein, this is a mistaken conception of what philosophers can expect to accomplish: "Philosophers constantly see the method of science before their eyes, and are irresistibly tempted to ask and answer questions in the way science does. This tendency is the real source of metaphysics, and leads the philosopher into complete darkness" (BB, p. 18).

Wittgenstein's student and lifelong friend Maurice Drury, who studied medicine and later specialized in psychiatry, tells this story about an oral examination he once took in physiology. Drury's examiner explained to him that Sir Arthur Keith had once told him that the reason the spleen drained into the portal system was of the utmost importance. But Keith had never explained just why this was so important. "Now," the examiner asked Drury, "can you tell me?" Drury confessed that he could not. The examiner went on. "Do you think there must be a significance, or an explanation? As I see it, there are two sorts of people: one who sees a bird sitting on a telegraph wire and says to himself, 'Why is that bird sitting just there?' The other man says, 'Damn it all, the bird has to sit somewhere.'"[11]

This is just the kind of story Wittgenstein might have told. We are led into philosophical confusion, thought Wittgenstein, by seeking explanations at times when we should instead simply stop and say, "This is how things are." At some point, reasons give out; explanations come to an end. As he famously put it, "If I have exhausted the justifications I have reached bedrock, and my spade is turned" (PI § 217). The urge to discover ultimate explanations, to seek out final justifications for a philosophical position, is what leads us to the sort of foundational, metaphysical philosophy that the later Wittgenstein rejected: "This is connected, I believe, with our wrongly expecting an explanation, when the solution of the difficulty is a description, if we give it the right place in our considerations. If we dwell upon it, and do not try to get beyond it. . . . The difficulty here is: to stop" (Z § 314).

For Wittgenstein, philosophical problems arise because we do not command a clear view of our language. "A philosophical problem has the form: I don't know my way around" (PI § 123). Philosophy does not add to what we know about the world; rather, it untangles the knots in our thinking caused by the way we represent the world through language (Z §

452). Solutions to our philosophical problems are not found through the construction of metaphysical foundations for our practices, or by building ever more elaborate philosophical theories, or even by constructing rules to restrict the way we are allowed to use words (PI § 133). Philosophical problems "are solved, not by giving new information, but by arranging what we have always known. Philosophy is a battle against the bewitchment of our intelligence by means of language" (PI § 109).

Wittgenstein's view of the role and purpose of philosophy clearly runs against the grain of quite a lot of contemporary bioethics scholarship, much of which operates explicitly within the domain of ethical theory.[12] In at least one regard, however, Wittgensteinians have some company. Feminists, clinical ethicists, medical anthropologists, literary scholars, and, perhaps most notably, the new casuists, who take their bearings in part from the distinguished work of one of Wittgenstein's students, Stephen Toulmin, all have criticized ethical theorists for preferring the thin air of moral theory to the thick, particular contexts in which moral problems are situated.[13] As a result, many bioethicists have been refocusing their attention lately on the concrete particularities of moral problems and the way they are described—for instance, through narrative, ethnography, and clinical case studies.

Yet it is also probably fair to say that many philosophers feel these particularist critiques as threatening, not simply because of what they believe philosophy is about, but because of what they believe morality is about. For many philosophers, if morality has no theoretical architecture, no foundation on which to stand, then our moral judgments and practices in particular cases seem to rest on nothing at all. As Margaret Olivia Little puts it, the "presumption that there are moral principles capable of codifying morality is deeply tied to the presumption that there had better be such principles"—or else our moral judgments seem to become dangerously idiosyncratic.[14]

For Wittgenstein, judgment need not be codified by rules or principles. Indeed, it cannot be codified, at least not entirely: "Our rules leave loopholes open, and the practice has to speak for itself" (OC § 139). Because rules are uneliminably ambiguous, as James Lindemann Nelson notes, their ambiguity is resolved by initiating learners into a practice. Learning how to make sound clinical judgments or sound moral judgments requires that "a community of some sort initiate a learner into the various ways of seeing and of going on to which that group regularly resorts." [15]

On a practical level, a rejection of moral theory is worrying to many bioethicists simply because it seems to rule out much of what they see as their most important job: providing theoretical order to the array of moral problems that practitioners face and developing a philosophical system for dealing with them.[16] This could hardly be more different from the later Wittgenstein's conception of philosophy. Wittgenstein once remarked that while other philosophers wanted to show that things that looked different are really the same, his aim was to show that things that look the same are really different. Wittgenstein was wary of the seductive appeal of theories and warned against explanatory schemes that bring a wide array of seemingly disparate concepts together. Wittgenstein considered using as an epigraph for the *Philosophical Investigations* Bishop Butler's remark, "Everything is what it is and not another thing." [17]

Perhaps even more worrying to bioethicists is the fear that by doing away with ethical theory, we also do away with the notion of the bioethicist as a kind of professional expert. If bioethicists cannot even claim any privileged access to theory, traditionally the domain philosophers have felt most comfortable calling their own, then any claims to expertise appear to be effectively squashed. So much the better, some might say; if expertise means a claim to authority or knowledge, then perhaps we are better off with bioethics purged of experts. But what if getting rid of bioethics experts means taking seriously the idea that bioethicists can claim no privileged position whatsoever—no special authority to guide a moral conversation, no special position from which to evaluate moral arguments, no claim to authority by virtue of having read and written about bioethical issues? Paul Johnston suggests that bioethics without experts would mean that there would be no difference between a bioethicist's speaking as a professional and her speaking as a layperson, and that the bioethicist as teacher would have no right to claim that her assessment of ethical issues was superior to that of her students. Johnston writes that in such circumstances, "the bioethicist does not have a unique perspective to offer doctors, nor does she know something they do not. If she offers a direction, a vision or warnings, she can only do so in the same way as anyone else." [18]

Perhaps then the work of the (so-called) bioethicist would look less like the work of a moral authority and more like the work of a philosopher, sorting out sense from nonsense, clearing a way through linguistic thickets, battling against the bewitchment of our intelligence by language. Yet there may be reason to worry about this view of bioethics as well. That is, if the philosopher is limited to a kind of conceptual analysis—if philosophy truly "leaves everything as it is"—then we seem to be left without the philosophical tools to press for radical moral change. For many philosophers, especially those who see cruelty or injustice in our current practices, this picture of moral philosophy seems dangerously conservative. Does Wittgenstein's view of philosophy have room for moral reformers? If, like Cora Diamond and David DeGrazia, you see the way we treat nonhuman animals as morally unjustifiable, then how, on Wittgenstein's account, do you move toward changing these practices?

One option, of course, is to reject Wittgenstein's view of philosophy. DeGrazia, for example, concedes that passages from the *Investigations* indicate that Wittgenstein was probably opposed to any sort of ethical theory, but he regards Wittgenstein's position as wrongheaded and dogmatic. For DeGrazia, Wittgenstein's anti-theoretical stance is a mistake. Yet even taking Wittgenstein's anti-theory seriously, he says, does not necessarily rule out radical moral reform.[19] DeGrazia reads Wittgenstein's stance against theory as an argument against the effort to evaluate our ethical and epistemological practices from a privileged, authoritative moral framework. But this stance, suggests DeGrazia, is still compatible with arguments for moral reform. That is, even if we admit that we cannot escape our own moral framework and achieve a God's-eye view of the world, we can still make the linguistic and conceptual moves allowed within that framework. For us, these moves might include things such as plausible reason-giving, fidelity to truth, and other conceptual commitments that can be marshaled in support of moral change.

Held Captive by Pictures

Wittgenstein himself held a somewhat different view of the kinds of moves one can make in ethics. As Cora Diamond points out, Wittgenstein suggested that reasons in ethics are like reasons in aesthetics.[20] These reasons, Wittgenstein says, are like "further descriptions." For example, you can help a person understand Brahms by showing him lots of compositions by Brahms, or perhaps by comparing Brahms to a certain kind of writer. But these reasons do not rationally compel agreement; the other person is not logically forced to adopt either your view or any other. Rather, they tell the listener, "Look at it this way." At best, your reasons might help the person see what you see. But if they do not, then that is the end of the matter. The analogy Wittgenstein said he had in mind was a discussion in a court of law, where you try to clear up the circumstances of the case, hoping that what you say will appeal to the judge. You draw his attention to certain things, or place certain things side by side, and by so doing you hope to change the way he sees the case. Diamond, for example, wants to place our attitudes and practices toward animals next to our attitudes and practices toward vulnerable humans, with the hope of changing the way we see and behave toward animals. Wittgenstein describes something like this kind of comparison in *Zettel*: "I wanted to put this picture before your eyes, and your acceptance of this picture consists in your being inclined to regard a given case differently; that is, to compare it with this series of pictures. I have changed your way of seeing" (z § 461).

The idea that Wittgenstein is putting forward here, that the philosopher can change the way we see certain practices or concepts by suggesting different pictures to us, is one that surfaces regularly in Wittgenstein's writing.[21] Wittgenstein thought that our language presents certain grammatical pictures to us, which then go on to dominate or constrain our thinking in subtle ways. The task for the philosopher is to liberate us from the grip of these pictures. As James Edwards says, Wittgenstein's technique "aims less at changing specific beliefs and more at getting oneself to see something differently, to pick up the stick at the other end."[22]

One way that grammatical pictures can mislead us is by giving us a narrow and constricted view of our language and the way certain words and concepts are used, causing us to overlook or forget other possible uses. Sometimes the pictures in front of us are merely too simple and gloss over the richness and variety of actual practices and attitudes. Larry Churchill thinks this is the case with our pictures of what constitutes a good way to die, for instance, and so he wants to emphasize the diversity in values held by people as they seek a good death.[23]

But grammatical pictures can also carry more insidious dangers. They can fool us into thinking that we have apprehended the way things really are, when in fact we are only seeing what our language, our concepts, present to us. As Wittgenstein puts it, "We predicate of the thing what lies in the method of representing it" (PI § 104). That is, we are led into believing that such pictures somehow represent the brute, metaphysical facts on which our language is built.[24] Often we do not even realize the extent to which our thinking is dominated by such a picture. This is what Wittgenstein is getting at in the *Investigations* when he says: "A picture held us captive. And we could not get outside it, for it lay in our language and language seemed to repeat it to us inexorably" (PI § 115).

One especially pervasive grammatical picture is that of the mind as a kind of private, inner theater to which only the thinker has access. This picture exerts a powerful grip on the way we think about the brain and human behavior. Nowhere is this more apparent than in clinical neurology and neurosurgery, where patients with neurological damage challenge our ordinary assumptions about mental life. For example, Grant Gillett is concerned with the widely held notion that patients in a persistent vegetative state may nonetheless have a conscious mental life, in spite of their failing to show any outward signs of consciousness. Many people worry, for instance, that such patients can feel pain, even though they show no outward signs of it. This worry is at least partly a result of the Cartesian picture of a mind whose internal workings are accessible to others only through language. The task for philosophers like Gillett, then, is to deliver us from the conclusions about vegetative patients toward which such a picture of the mind leads.[25]

Wittgenstein's Sensibility

Stephen Toulmin once told me that trying to put together a book on bioethics and Wittgenstein would be like trying to put together a book on bioethics and William Blake. The comparison to Blake is especially apt, because it points toward a discordance that goes beyond temporal and cultural differences. It indicates a difference in sensibility that makes a book on Wittgenstein and bioethics even more eccentric than a book connecting bioethics to Kant, or Hume, or even Plato. Remember that Wittgenstein was a man who, born into one of Austria's most prosperous and cultured families, gave away the entirety of his inherited wealth; who, after writing one of the twentieth century's most important works of philosophy, left university life to become a schoolteacher for six years in an Austrian peasant village; who was known to his comrades in World War I as "the man with the gospels" after he discovered Tolstoy's *The Gospel in Brief* in a Galicia bookstore and then carried it with him at all times throughout the war; who spent extended periods of his life writing philosophy alone in isolated parts of Norway and Ireland; who, instead of discussing his philosophical work with his positivist admirers in the Vienna Circle, is said to have insisted on reading them poetry; and who, in the early part of his life, as Bertrand Russell wrote with astonishment, "reads people like Kierkegaard and Angelus Silesius, and he seriously contemplates becoming a monk." [26] More than anything else, it is Wittgenstein's austere, intensely spiritual, almost evangelical sensibility that drives against the tenor and spirit of contemporary bioethics.

This helps us see why it would sound odd, even bizarre, to apply the term "ethicist" to Wittgenstein, even though ethics was clearly at the heart of his life and philosophy. Wittgenstein's sensibility is profoundly at odds with the overtones of professionalism that the word "ethicist" carries: the managerial ethos of the professions, their instrumental attitude toward knowledge, perhaps even the lingering air of bourgeois financial comfort with which they are associated. "Ethicist" implies a kind of role playing, the habitation of a professional persona, that seems alien to Wittgenstein's vision of philosophy. For Wittgenstein, philosophy was not a professional enterprise but a personal quest. It was an activity upon which everything was staked.

According to Maurice Drury, Wittgenstein once shocked G. E. Moore by insisting that in philosophy, character is more important than intelligence. [27] Why Moore was taken aback is not hard to see; if philosophy (of all things) is not a matter of intellect, then what is? Yet Wittgenstein's insistence on the importance of character for philosophy is consistent with his vision of what sound philosophy involves. Wittgenstein's philosophy involves a kind of humility; it is not just about knowing what to say, but about knowing what not to say. The danger of academic life, thought Wittgenstein, is that we are encouraged to go on talking even when we know in our hearts that we have nothing valuable to say. [28] Thus we must always be vigilant in resisting the temptation to say more than we really know. This vigilance amounts to a kind of mental asceticism, less a matter of the intellect than of the will. It is not an exaggeration to say, as Knut Erik Tranöy does, that Wittgenstein's intellectual genius was a product of his moral seriousness and sincerity.[29] Whatever else might be said of Wittgenstein as a philosopher, he cannot be accused of mere cleverness or intellectual gamesmanship. It is difficult to imagine Wittgenstein making an argument simply for the sake of argument, or writing philosophy to secure academic advancement. Drury once remarked to Wittgenstein that one of his friends had abandoned his postgraduate studies on the grounds that he had nothing original to say. To this Wittgenstein replied, "For that action alone they should give him his Ph.D." [30]

Nor does Wittgenstein's view of ethics mesh comfortably with the views of ethics implicit in contemporary bioethics scholarship. For Wittgenstein, ethics was not simply about our duties and obligations to others, or even about questions of right conduct and good character; it was about "the meaning of life," about "what makes life worth living" (LE, p. 5). This is a vision of ethics at once broader and deeper than that of most bioethics scholarship, which generally concerns itself with more limited questions about what kinds of practices are ethically (or legally) permitted. What Wittgenstein criticizes in philosophers like Russell might also be said of much of contemporary bioethics: "Some philosophers (or whatever you like to call them) suffer from what might be called 'loss of problems.' Then everything seems quite simple to them, no deep problems seem to exist any more, the world becomes broad and flat and loses all depth, and what they write becomes immeasurably shallow and trivial" (z § 456).

This "loss of problems," or absence of wonder, is something that Wittgenstein felt not only in certain philosophers but in the work of novelists and musicians. He once told Drury, "I am not a religious man but I cannot help seeing every problem from a religious point of view." [31] He rebuffed his positivist admirers and was drawn instead to writers like Kierkegaard, Blake, Tolstoy, and Dostoyevsky. His own "Lecture on Ethics" conveys above all a sense of awe, even reverence, toward the ethical. "I can only express my feeling by the metaphor, that, if a man could write a book on ethics which was really a book on ethics, this book would, with an explosion, destroy all the other books in the world" (LE, p. 10.)

For Wittgenstein, the appearance of a philosophical problem is a sign that something has gone wrong, a pathology that the philosopher must diagnose and treat. He sometimes referred to philosophical confusion as a disease for which his method of philosophizing is a cure (PI § 133, z § 382). "The philosopher's treatment of a question is like the treatment of an illness" (PI § 255). Philosophy is not just a way of sorting our way through vexed questions, however; it is a treatment of the self, a therapeutic correction for the disordered sensibilities to which our language and practices lead. What Stanley Hauerwas says of Christian discourse could equally well be said of Wittgenstein's philosophy: that it is "not a set of beliefs aimed at making our lives more coherent; it is a constitutive set of skills that requires the transformation of the self to rightly see the world." [32] Indeed, part of what Wittgenstein found to admire in Christianity was the recognition that, as he put it, "sound doctrines are useless" (CV, p. 53). Christianity, he wrote, "says that wisdom is all cold; and that you can no more use it for setting your life to rights than you can forge iron when it is cold" (CV, p. 53).

But it is important to realize that the illnesses with which Wittgenstein is concerned are not peculiar to patients who are inclined toward unnecessary philosophizing. They are the kinds of illnesses to which any one of us is susceptible, because they are embedded in our language and our form of life. For Wittgenstein, language does not stand on its own; it is an activity tied to a culture's particular ways of seeing and relating to the world: "t[o] imagine a language is to imagine a form of life" (PI § 19). The philosophical illnesses to which we are susceptible are those endemic to particular cultures at particular times and to the sensibilities they produce: "the philosopher is the man who must cure himself of many sicknesses of the understanding before arriving at the notion of the sound human understanding. If in the midst of life we are in death, so in sanity we are surrounded by madness" (RFM, p. 157).

Just how we might overcome this kind of illness is what Wittgenstein is getting at when he writes, "In philosophizing we may not *terminate* a disease of thought. It must run its natural course, and *slow* cure is all important. (That is why mathematicians are such bad philosophers)" (z § 382). Diseases of thought are rarely the product of of a single misguided analogy or false grammatical picture. More often they are the result of a large web of interconnected conceptual structures that cannot be easily dismantled. They are rooted deeply in our language and our form of life, and their cure is more like treating psychopathology with psychotherapy than like treating a respiratory infection with an antibiotic. This sort of treatment takes time.

In the end, a different sensibility may be the most important thing that a careful reading of Wittgenstein has to offer bioethics: not a philosophical doctrine, but an attitude toward the world and our place in it. Yet for Wittgenstein, achieving a different sensibility is not solely a matter of what we do or decide. It is a product of a language and a form of life. "The sickness of a time is cured by an alteration in the mode of life of human beings, and it was possible for the sickness of philosophical problems to get cured only through a changed mode of thought and of life, not through a medicine invented by an individual" (RFM, p. 57). As James Edwards puts it, "The sound human understanding is not our achievement. It is possible for us because a form of life is, and that possibility is not the result of our will. . . . The sound human understanding is an occasion of grace, not heroic triumph."[33] Given the state of U.S. medicine and bioethics, this does not leave us with an optimistic conclusion. In a time when medical practice is driven by free-market capitalism, utilitarian ethics, and a Whiggish optimism about the power of technology, it is hard to disagree with Wittgenstein's pessimism about his work: "It is not impossible that it should fall to the lot of this work, in its poverty and the darkness of its time, to bring light into one brain or another—but of course, it is not likely" (PI vi).

Notes

Parts of this introduction are taken from my review of Maurice Drury's *The Danger of Words and Other Writings on Wittgenstein, Hastings Center Report* 28, no. 1 (1998): 38–40.

1 Norman Malcolm, *Ludwig Witttgenstein: A Memoir*, 2d ed. (Oxford: Oxford University Press, 1984), p. 98.

2 Karl Britton, "Portrait of a Philosopher," in *Ludwig Wittgenstein: The Man and His Philosophy*, ed. K. T. Fann (New York: Dell, 1967), p. 62.

3 M. O'C. Drury, *The Danger of Words and Writings on Wittgenstein*, ed. David Berman, Michael Fitzgerald, and John Hayes (Bristol: Thoemmes Press, 1996), p. 123.

4   Ray Monk, *Ludwig Wittgenstein: The Duty of Genius* (London: Vintage Books, 1991), p. 323.

5   Monk, *Ludwig Wittgenstein*, p. 294.

6   John Hayes, "Wittgenstein's 'Pupil': The Writings of Maurice O'Connor Drury," in M. O'C. Drury, *The Danger of Words and Writings on Wittgenstein*, ed. David Berman, Michael Fitzgerald, and John Hayes (Bristol: Thoemmes Press, 1996), p. xx.

7   Monk, *Ludwig Wittgenstein*, p. 425.

8   Malcolm, *Ludwig Wittgenstein: A Memoir*, p. 28.

9   Ibid., p. 94.

10   James C. Edwards, *Ethics without Philosophy: Wittgenstein and the Moral Life* (Tampa: University of South Florida Press, 1985), p. 150.

11   Drury, *The Danger of Words*, p. xi.

12   See, e.g., (among many others): K. D. Clouser and B. Gert, "A Critique of Principlism," *Journal of Medicine and Philosophy* 15 (1990): 219–36, and "Morality versus Principlism," in *Principles of Health Care Ethics*, ed. Raanan Gillon (Chichester: John Wiley and Sons, 1994), pp. 251–66; Robert Veatch, *A Theory of Medical Ethics* (New York: Basic Books, 1981); Tom Beauchamp and James Childress, *Principles of Biomedical Ethics*, 4th ed. (New York: Oxford University Press, 1994); and Raanan Gillon, preface, "Medical Ethics and the Four Principles," in *Principles of Health Care Ethics*, ed. Raanan Gillon (Chichester: John Wiley and Sons, 1994), pp. xxi–xxxi.

13   For a representative sample of these critiques, see Albert R. Jonsen and Stephen Toulmin, *The Abuse of Casuistry: A History of Moral Reasoning* (Berkeley: University of California Press, 1988); Edwin R. DuBose, Ronald P. Hamel, and Laurence J. O'Connell, eds., *A Matter of Principles? Ferment in U.S. Bioethics* (Valley Forge, Pa.: Trinity Press International, 1994); Hilde Lindemann Nelson, ed., *Stories and Their Limits: Narrative Approaches to Bioethics* (New York: Routledge, 1997); Susan M. Wolf, ed., *Feminism and Bioethics: Beyond Reproduction* (New York: Oxford University Press, 1996); Kathryn Montgomery Hunter, *Doctors' Stories: The Narrative Structure of Medical Knowledge* (Princeton: Princeton University Press, 1991); Carl Elliott, "Where Ethics Comes from and What to Do about It," *Hastings Center Report* 22, no. 4 (1992): 28–35; Anne Hunsaker Hawkins, *Reconstructing Illness: Studies in Pathography* (West Lafayette, Ind.: Purdue University Press, 1993); Ronald Carson, "Interpretive Bioethics: The Way of Discernment," *Theoretical Medicine* 11 (1990): 51–59; Carl Elliott, "Hedgehogs and Hermaphrodites: Towards a More Anthropological Bioethics," in *Philosophy of Medicine and Bioethics: A Twenty-Year Retrospective and Critical Appraisal*, ed. Ronald A. Carson and Chester R. Burns (Dordrecht: Kluwer Academic, 1997), pp. 197–211; and Howard Brody, *Stories of Sickness* (New Haven: Yale University Press, 1987).

14   See Little's essay in this volume, "Wittgensteinian Lessons on Moral Particularism."

15   See James L. Nelson's essay in this volume, " 'Unlike Calculating Rules'? Clinical Judgment, Formalized Decison Making, and Wittgenstein."

16   See Little, "Wittgensteinian Lessons."

17   Paul Johnston, *Wittgenstein and Moral Philosophy* (London: Routledge, 1989), p. 14.

18 See Paul Johnston's essay in this volume, "Bioethics, Wisdom, and Expertise."

19 See DeGrazia's essay in this volume, "Why Wittgenstein's Philosophy Should Not Prevent Us from Taking Animals Seriously."

20 See Diamond's essay in this volume, "Injustice and Animals." The original source of Wittgenstein's remarks is G. E. Moore, "Wittgenstein's Lectures in 1930–33," reprinted in *Classics of Analytic Philosophy*, ed. Robert R. Ammerman (New York: McGraw-Hill, 1965), p. 278. See also James Edwards's discussion in *Ethics without Philosophy*, p. 129.

21 Edwards has an excellent discussion of this in *Ethics without Philosophy*, pp. 117–25.

22 See Edwards's essay in this volume, "Religion, Superstition, and Medicine."

23 See Churchill's essay in this volume, "Patient Multiplicity, Medical Rituals, and Good Dying: Some Wittgensteinian Observations."

24 Edwards, *Ethics without Philosophy*, p. 118.

25 See Gillett's essay in this volume, "Wittgenstein's Startling Claim: Consciousness and the Persistent Vegetative State."

26 Quoted in Edwards, *Ethics without Philosophy*, p. 24. Russell wrote this in a letter to Lady Ottoline Morrell in 1919 that is included in Ludwig Wittgenstein, *Letters to Russell, Keynes, and Moore*, ed. G. H. von Wright (Oxford: Basil Blackwell, 1967), p. 82.

27 Drury, *The Danger of Words*, "The 1967 Dublin Lecture on Wittgenstein," p. 8.

28 See M. O'C. Drury's contribution to "A Symposium: Assessments of the Man and the Philosopher," in K. T. Fann, *Ludwig Wittgenstein: The Man and His Philosophy* (New York: Dell, 1967), p. 69.

29 See Knut Erik Tranöy's essay in this volume, "Wittgenstein: Personality, Philosophy, Ethics."

30 Drury in Fann, *Ludwig Wittgenstein: The Man and his Philosophy*, p. 69.

31 Drury, *The Danger of Words*, "Some Notes on Conversations with Wittgenstein," p. 79.

32 Stanley Hauerwas, *Dispatches from the Front: Theological Engagements with the Secular* (Durham, N.C.: Duke University Press, 1994), p. 7.

33 Edwards, *Ethics without Philosophy*, p. 255.

2 Religion, Superstition, and Medicine
James C. Edwards

Is religious faith just a kind of superstition? The question seems absurd, because the identification it offers is so crassly reductive. Surely there is all the difference in the world between someone's being afraid of breaking a mirror, or carefully stepping over the cracks in the sidewalk, or rubbing a rabbit's foot before the big game, and, on the other hand, the Snake Dance performed every second year in the Hopi pueblos, or the skeins of close reasoning deployed in Calvin's *Institutes*, or Mother Teresa praying over the dying she has collected from the streets of Calcutta.

True enough, but the question will not go away so easily. What do we make of practices, especially ritual practices, that seem to rest upon, or at least to imply, beliefs that are manifestly false? In particular, what do we make of those ritual practices that seem to advert to causal powers not of the sort with which our established natural and social sciences deal, practices that trace our well-being to the identification and propitiation of those mysterious and sometimes personal powers? Why does one work so hard to avoid breaking a mirror? Because one does not want to release the seven years of bad luck held in check by its shiny wholeness. Why does the climber take care not to insult Mount Nyangani, the highest in Zimbabwe? Because she does not want to offend the spirits resident there and thus put her life in danger.[1]

But we end-of-century, Western intellectuals do not believe in those seven years of bad luck encapsulated in the unbroken mirror, nor do we—most of us—believe in ghostly spirits guarding an African mountain from scoffers. Is that not what leads us, or inclines us, to call such things superstitious?[2] We cannot understand the practice (mirror guarding, mountain reverencing) apart from the beliefs we (or they, the persons whose practice it is) cite as its ground and explanation; yet those beliefs are manifestly and uncontroversially false. Moreover, we think it is easy for us to see what has given rise, and continuing power, to those obviously false beliefs: our need to console ourselves, to flatter ourselves with the assurance of control we do not really have, and to hide from ourselves our frightening vulnerability to accident and malice.

Given this model of superstition, it is tempting to apply it to one's own religious heritage. If it is a kind of superstition to treat Mount Nyangani as the home of dangerous spirits that must not be offended, then why should we not come to see the Eucharist in the same light?[3] That will at first be more difficult, of course, since the beliefs bound up in the celebration of the Eucharist are for most readers of this essay much closer to home than the beliefs bound up in the reverence for Mount Nyangani, and our nearness, both cultural and individual, to those former beliefs makes us reluctant to view them with a sufficiently cold eye. But that is a temporary impediment. Soon enough, under pressure of our philosophically enhanced will to truth, we will be able to look at our own lives anthropologically, as if they were the lives of strangers; and at that point we too will see that the beliefs (e.g., transubstantiation) that underlie our (current or former) religious practices are manifestly false. Not only that: those beliefs too can be plausibly represented as comforting and deceptive fantasies, as attempts to camouflage our frightening and humiliating subjection to circumstance. We will thus be able to see our own religious superstitions for what they are.

But most of you would no doubt agree with me that this way of thinking is sophomoric; and, to fall into an appropriately sophomoric idiom, that call is a no-brainer. It is not hard for us to be convinced that a crude comparison between religion and superstition is ill conceived; what is hard is to see exactly what is wrong with it. This essay sketches an answer to that question. I want to show, quite briefly, how religion lends itself to a superstitious interpretation in the hands of philosophers, even some classic pragmatist and neopragmatist philosophers; and then to show, even more briefly, how it might avoid that fate. I will conclude with some remarks about medicine—like religion, a practice where our susceptibility to superstition is all too common.

———

I will begin with a story for us to think about. When I was in my early twenties and just beginning to study philosophy, I knew a woman, then in her sixties, whose husband died suddenly of a heart attack. This woman (Mrs. Darlington, I will call her) was the particular friend of my wife's mother, and my mother-in-law was worried about her friend's odd behavior in the months just after the funeral. In the late afternoon of every day, rain or shine, Mrs. Darlington went to her husband's grave and spoke, out loud, to him. She would sit down on the ground—being quite fastidious, she always brought a blanket to spread out on the grass—and cry her eyes out. Then, having recovered herself somewhat, she would tell Mr. Darlington what had happened at home that day, filling him in on all the quotidian details of her life. Sometimes she would ask him questions (and wait for the answers), and sometimes she put to him various grievances his death had left her with. Sometimes she would even bring things to show him, such as recent photographs of the grandchildren or freshly gathered produce from the garden she and he had worked in side by side. Her speech was perfectly audible, animated in fact, and was noticed with some consternation by other visitors to the graveyard, who reported the matter to her family and friends. Mrs. Darlington kept up these visits for about a year, quite regularly, and then they simply stopped. She never talked to her friends or family about her trips to the cemetery (except to mention them in passing, as if they were to be expected), and although those who cared about her were sometimes fearful of where her behavior might be leading, they never worked up the courage to intervene. Then she stopped going, as I have said. She lived for a good number of years after that, missing her husband, as was natural, but visiting his grave, and silently now, only on the anniversary of his death.

What are we to make of these events? I will tell you what I made of them then, in the first full flush of my graduate-school wisdom. I saw Mrs. Darlington as pathetically superstitious. I thought she had, and was acting upon, several false but comforting beliefs, some of which she had had for a long time and could be attributed to her religion, others of which were recently acquired and were explicable as the result of the shock she had received at Mr. Darlington's sudden death. I thought she believed, as was natural for her sort of Christian, that her husband had survived his death and was then alive and conscious in the heavenly presence of his Lord;

and I further thought she believed that by going out to his grave she somehow could actually make him, now resident in heaven, hear her as she talked to him. (I am firmly convinced she herself would have said she believed those very things, if someone had been so insensitive as to quiz her about her behavior.) And both those beliefs I took (and still take) to be false: not only false, but obviously false, given the utter lack of convincing evidence that could be brought forth in their behalf. Is that not what superstition is—acting on comforting beliefs for which all good evidence is manifestly lacking?

Today that strikes me as a callow and (on my side, not hers) superstitious way of thinking about Mrs. Darlington's graveside visits and harangues. Now I would be inclined to say, that was just the way her grief took her. That was just the way she lived her grief at Mr. Darlington's death; and when she had lived her grief out, or most of it anyway, she stopped going to the cemetery, and she stopped talking out loud to her dead husband. Today I would not be inclined to think that her disconcerting behavior was something that needed to be, or could be, explained by reference to her beliefs, superstitious or otherwise. Rather, I believe it stemmed from something prior to any belief she or we might dream up in an attempt to explain it. Grief at the death of one's mate is, after all, a part of the natural history of human beings; and although it explains some of what we sometimes do, it does not need explaining itself, any more than breathing or thinking does. (We certainly do not explain our breathing or our thinking by reference to any beliefs we have about them, beliefs about their "usefulness" or "efficacy," for example.) Grief—or thinking—is just there (sometimes); it is part of the weave of a human life. It is what Wittgenstein might call a "proto-phenomenon." [4] Grief is part of our natural history, part of the bedrock at which our philosophical spades, digging up explanations in terms of something hidden (such as a belief, a soul, or a kinesthetic sensation), are simply turned. We can say of grief: it happens to us; it belongs to the form of life we are. "It is there—like our life." [5] And I want to say: in Mrs. Darlington's eccentric behavior we see grief itself. Not the "expression" of grief, or her "dealing with" grief, or her "grief-work": we see grief itself. It is visible to us directly, as the thing it is, and as such it rests on nothing further, such as beliefs or convictions or experiences. This is human grief, pure and simple, *die Sache selbst*.

Of course it is hard for us to maintain such an anti-philosophical stance toward things. We feel a great need to explain what we see, and to explain it in a particular way, by bringing out something hidden that accounts for what stands before us in plain view. Much of the time, with ourselves and with other people, we follow a particular variant of that plausible strategy of revelation: we try to account for behavior (especially puzzling behavior) by ascribing beliefs—beliefs that may be hidden even from the one who believes them—that count as motivating factors in actions, linguistic or otherwise. The (visible) actions, we think, express those (invisible) beliefs. We see others as having the same beliefs we have or can imagine ourselves having in their situation. And much of the time this hermeneutic strategy serves us well, as Quine and Davidson have demonstrated. But sometimes the strategy leads us wrong, too. Sometimes it is best—though difficult—simply to take a particular way of acting or reacting as a proto-phenomenon, to see it as a fundamental part of our natural history, in horizontal relation to other parts of that natural history rather than in vertical relation to something beyond or behind it. No hidden beliefs underlie such a proto-phenomenon as its explanation or justification (though of course there may be hidden causes for its occurrence); it is just there, prior to all belief and ratiocination, prior to all such vertical explanation. If we were to come upon a stranger weeping beside the lifeless body of his companion, would it be useful (or even sensible) to say that he must believe that it is a bad thing his friend is dead?[6] Wouldn't it be better to say, "Being taken by this kind of grief is part of what it means to be human?" Such grief rests on and expresses no explanatory or justificatory beliefs. It is as natural and as primitive (in terms of explanation) as breathing.

If someone had asked Mrs. Darlington about her visits to her husband's grave, she would no doubt have responded by recounting beliefs that would "explain" what she was doing ("my darling Lee is alive in heaven and can hear what I say to him"). Given her religious background, she had a stock of such beliefs ready to hand, and no doubt she would have used them (and perhaps actually did, if she herself had any middle-of-the-night qualms about what she was doing). No doubt such beliefs were sincerely held, too, though it is also entirely likely she would have felt a bit strange about having to trot them out. Given the fact that she would have produced such beliefs, and would have produced them as an explanation of why she was doing what she was doing, it is entirely natural for us to take her at her word and thus to take these beliefs as the always already present explaining or justifying ground for the behavior we are puzzled about. It is natural, that is, for us to see the behavior as the natural and inevitable expression of the prior beliefs: if one believes x, then one will do y; if one is doing y, it must be because one believes x.

But I am trying to suggest that this would be a mistake. I am trying to suggest that the distressing behavior is not the expression of anything; that it is her grief itself, taking her in a particular way. I am further suggesting that her beliefs—that is, what she says when feeling called upon to explain and to justify (even if only to herself) her behavior—do not really play the role they seem (and announce themselves) to play. Rather than being prior explanatory or justificatory conditions of the behavior, conditions now brought to consciousness and known to be such, the beliefs are retrospective interpretations of the behavior. As such efforts at interpretation, they of course avail themselves of other interpretations and styles of interpretation already available in the culture (e.g., Christianity, belief in an afterlife, justification of action by appeal to underlying beliefs, explaining the manifest by invoking the hidden, and so forth), just as any of us tries to use the tools currently available to us when we have a job to do. And that is what Mrs. Darlington might have done. If she had felt some need to account for the way her grief was taking her (putting it crudely, to account for her need to sit by her husband's grave and to talk to him, and to account for the satisfaction those things brought her), she would account for it in the easiest way she could, by invoking those theological beliefs available to her as a middle-of-the-road Methodist Christian, beliefs she now retrospectively assumes to be, and asserts to be, underlying her visits to the graveyard.

In these reflections on Mrs. Darlington I have been following the line run by Wittgenstein in his remarks on Frazer's anthropology.[7] Frazer does see religion (especially the religion of so-called primitive peoples or rude societies) as a kind of superstition. When confronted with strange and impressive rituals and practices (e.g., the killing of the priest-king of Nemi), Frazer—like a good Quinean or Davidsonian linguist encountering utterances in an unfamiliar language—explains those actions as the expressions of beliefs held by the participants: beliefs we share with them, or can at least imagine ourselves sharing with them. (He is helped in this endeavor by the accounts of the practices given to the investigator by the participants themselves, accounts that do refer to various beliefs in connection with the practices.) In particular, Frazer explains "primitive" rituals and practices by construing them as expressing beliefs that compete with the claims of our contemporary natural sciences, and thus beliefs that we can see to be errors, to be manifestly and uncontroversially false. In this way the practices of the "rude societies" can be simultaneously understood and dismissed: understood in terms of the faulty natural science of our ancestors, dismissed for those very faults.

Here is Frazer under full sail:

> But reflection and enquiry should satisfy us that to our predecessors we are indebted for much of what we thought most our own, and that their errors were not willful extravagances or the ravings of insanity, but simply hypotheses, justifiable as such at the time they were propounded, but which a fuller experience has proved to be inadequate. It is only by the successive testing of hypotheses and rejection of the false that truth is at last elicited. After all, what we call truth is only the hypothesis which is found to work best. Therefore in reviewing the opinions and practices of ruder ages and races we shall do well to look with leniency upon their errors as inevitable slips made in the search for truth, and to give them the indulgence which we ourselves may one day stand in need of: *cum excusatione itaque veteres audiendi sunt.*[8]

There is much here that Wittgenstein despises, but his primary point of attack is Frazer's assumption that the rituals and practices he is considering must be understood as being grounded in particular hypotheses affirmed by the participants, hypotheses "justifiable as such at the time they were propounded, but which a fuller experience has proved to be inadequate." The problem here is not (or not only) Frazer's patronizing assumption that our experience of the natural world is, in some relevant sense, "fuller" than that of the "primitives." Rather, Wittgenstein wants to disengage the significance of the practices under discussion from any "hypotheses" (or "opinions") they are supposed to express. He wants to be able to understand these practices, and their pathos for us and for their original participants, in another way: not vertically, as the visible expressions of something otherwise hidden, but horizontally, as standing on their own in relation to the life in which they visibly occur. He wants to present these practices as part of the natural history of human beings, in the same way that birdsong is part of the natural history of avifauna. Part of Frazer's obtuseness is his inability to break away from the picture that all deliberate and impressive human actions express beliefs, and that the significance of those actions is given always and only by the beliefs they express. But that inability is just an instance of a deeper one: the inability not to think of all the visible as given its significance by the invisible.

Put that way ("The visible is just the trace of the invisible, which grants it life and meaning"), Frazer's guiding assumption seems a paradigmatic case of the magical, superstitious thinking he claims to despise. In fact, his assumption is that part of magic we now call metaphysics: the Platonic need to explain the substance of "this" world by appeal to another one, a "true" one.[9] The meaning of the killing of the priest-king of Nemi cannot (we think) be in the killing itself or in its attendant circumstances and consequences; it must be found in the invisible, hidden world of the psyche, in the beliefs ("hypotheses" or "opinions") held by those who do the killing, beliefs that themselves refer to a hidden world of the spirit. What could be more magical and superstitious than the principle that the hidden, the invisible, is what always gives sense and life to the visible?

Frazer's thinking about the practices and rituals he discusses is just as primitive and superstitious as was my graduate-school thinking about Mrs. Darlington. We were both under the spell of metaphysics; both of us felt a need to account for something visibly strange and disconcerting by finding some hidden root for it and in that way making it manageable. We both looked for that hidden root in the mind, the beliefs, of the eccentric other who frightened us; and once those beliefs were (as we thought) identified, we knew where we stood. The authority of the practices—or, in this case, the lack of it—had been revealed by the discovery of their hidden sources. Their pathos could be endorsed or discounted, depending on the truth or falsity of the beliefs they expressed. Almost everyone who has tried to think philosophically about religion has fallen into the pattern of thinking at issue between Frazer and Wittgenstein here. And this is true even for pragmatist and neopragmatist thinkers, who ought to know better. William James, for instance, in "The Will to Believe," explicitly construes religion in terms of what is true to believe and even speaks of "the religious hypothesis" that he assumes to be at the heart of religious practice.[10] In *A Common Faith*, Dewey claims that the cause of religion's loss of power is the problem of belief—namely, that we can no longer believe those things we once did—and he looks about for other beliefs capable of underwriting the attitudes and ideals traditional religiousness was so good at inculcating.[11] And I have already pointed out how neopragmatists such as Quine and Davidson link their schemes of radical translation or radical interpretation to the beliefs (and the desires, of course)[12] held by the strangers they wish to understand. The Frazer model of practice as an expression of belief is readily apparent in all these cases.

In one way, of course, there is nothing wrong with that model. What other, better hermeneutic strategy is open to the linguist or anthropologist faced with radical difference? And what is wrong with talking about religion in terms of belief, truth, and hypothesis? Religious folks do so themselves, after all. I am certainly not suggesting, nor was Wittgenstein, that we try to approach the matter of religion in a purely behavioral way, whatever that might mean. That would be absurd. Religion, and, for that matter, superstition, are part of the weave of human life, and human beings can finally be understood only in terms of "the intentional stance."[13] We do have beliefs and desires, including religious ones, and we sometimes act on them. Sometimes, indeed, our actions, ritual and otherwise, can only be understood in terms of such beliefs and their expression. But—and this is an important qualification—there is something dangerous in this way of thinking, and Wittgenstein is on track in showing us what it is. Part of the danger is simple hasty generalization: if some human practices demand to be explained as the expression of beliefs and desires, then all do. (In Wittgenstein's view, such hasty generalization is much more common in human thinking than we are eager to admit. Much bad philosophy springs from this freshman-level fallacy.) But part of the danger Wittgenstein sees in Frazer has a more interesting root. This root, to put it crudely, is that we tend to fall under the spell of a particular picture of what a belief is, and of what it means to explain an action by reference to a belief. That is, we fall into thinking that the visible must be explained by reference to the invisible, and thus that the "beliefs" that "explain" the rituals and practices of religion are somehow hidden away in the psyches of those whose beliefs they are, psyches that must be plumbed so that they can reveal what the visible actions "mean." What is at issue between Wittgenstein and Frazer—and, if I am right, between Wittgenstein and the pragmatists and neopragmatists (at least sometimes)—is a large-scale picture of human practices and of the human beings whose practices they are. It is the picture of outside and inside, visible and invisible; it is the picture that the meaning of the former is always and only to be found in the latter. It is the magical picture of metaphysics: that the meaning and substance of "this" world is always and only to be found in one behind or above or below it, in a "true" world hidden from plain view, available only to those who have been taught to see it.

Nothing is wrong with that picture as such; surely there are times when the outer/inner, visible/invisible picture helps us to think fruitfully about human beings and our world.[14] The problem with the (grammatical) pictures that guide our thinking is always in their application to particular cases, not in the pictures themselves. And a major source of our misapplication of a given picture (e.g., applying it always and everywhere, without a proper regard for its peculiarities and limitations) is the lack of ready alternatives to it. If we are going astray in our wholesale application of the Frazer model of practice as an expression of belief to religious and magical rituals and practices, then what is the alternative picture Wittgenstein gestures toward?

A full, defensible, and defended answer to that question is still beyond us, but the outlines are easy to trace. Wittgenstein wants to help us to see religious and ritual practices as proto-phenomena, whose meaning is to be understood by hooking them up horizontally to the other parts of the life in which they occur rather than by seeing them as places where the "true world" of the spirit bubbles through into the ordinary.

> Burning in effigy. Kissing the picture of one's beloved. That is obviously not based on the belief that it will have some specific effect on the object which the picture represents. It aims at satisfaction and achieves it. Or rather: it aims at nothing at all; we just behave this way and then we feel satisfied.
>
> One could also kiss the name of one's beloved, and here it would be clear that the name was being used as a substitute.
>
> The same savage, who stabs a picture of his enemy apparently in order to kill him, really builds his hut out of wood and carves his arrow skillfully and not in effigy.[15]

Here Wittgenstein is construing the actions of the lover—kissing every night the picture of her beloved—as I came to construe the daily graveside harangues of Mrs. Darlington: as the thing itself, grief or love, not the expression of something (a belief) hidden in another realm (the mind). To understand the lover's actions it is not necessary to assume that she is acting on any particular convictions or desires, that she is, as Wittgenstein puts it, "aiming" at something in particular. No, all that is necessary is that one see the action—the kissing of the picture—in a proper relation to the weave of this woman's life: that it is a picture of A; that she loves A; that A is absent from her. And then one can see the action for what it is: an instance (not an expression) of her love for A. Of course, seeing it this way, as an instance of this woman's love for A, is to see it as a part of a human life, and of course, that is a life understood in terms of beliefs and desires. We could not understand this woman at all well if we did not take her to have, and to act upon, beliefs and desires that are her own. But that is different from seeing everything she does as being given its meaning and pathos by reference to the beliefs and desires of hers that it (allegedly) expresses. The intentional stance has its limits; some things are better seen as part of the natural history of human beings, as proto-phenomena.

I began this essay with a reductive comparison of religion to superstition, in the hope of seeing what makes that comparison—unsatisfactory as it no doubt is—nevertheless permanently tempting to philosophical thinkers. Following Wittgenstein's lead, I have suggested that the comparison takes its power from a particular picture of how human rituals and practices are finally to be understood: namely, as expressions of beliefs (or "hypotheses," or "opinions") and desires. If that picture of understanding is applied to religious rituals and practices, then the result is (sooner or later) what we see in *The Golden Bough*: those practices will come to seem rude and primitive attempts to do something we now do better, or even that we now see is not worth trying to do at all. In other words, they will come to seem superstitious—because ultimately magical—attempts to predict and control the natural and social vicissitudes that make our lives good or bad (or somewhere in between). I have been suggesting, again following Wittgenstein, that the superstition here is on the side of the philosopher who thinks the only way to explain what one sees is to advert to something that one does not see. Why should Frazer be so puzzled about the killing of the priest-king at Nemi? Because he assumes that the mean-

ing of this impressively disturbing act is hidden and mysterious, something as impressively disturbing as the act itself. Why was I, or my mother-
in-law, so provoked by Mrs. Darlington's visits to her husband's grave?
Because I, at least, assumed—or felt the need to assume—that the explanation for those eerie visits had to be found in some beliefs I could either
endorse or, much more likely, dismiss (thus dismissing the grief that so
frightened me in the first place: why was I so afraid of that honest suffering?). Puzzled or frightened by the visible, we seek solace in the invisible.
In that way, it is we who are the "savages," not those we condescend to call
by that name. Like the practice of philosophy, the practice of medicine is
rife with superstition, and with its consequences. As a counterpart to my
story about Mrs. Darlington, here is another, told to me by a friend who
was a part of it.

The matriarch of a large extended family was brought to the emergency
room of a teaching hospital; she was comatose and having serious respiratory difficulties. She was immediately put on a ventilator in order to stabi-
lize her labored and erratic breathing, and that procedure was successful:
on the machine she was getting plenty of oxygen, and her major organ
systems were functioning adequately. Nevertheless, within a day it was
clear to the house staff that the case was hopeless. Tests showed that the
patient had had a massive cerebral hemorrhage, or perhaps a series of
them, with extensive and irreparable damage to the brain. She would
never wake up; she was, in fact, brain dead, and thus, in this jurisdiction,
legally dead. The only thing keeping her body alive was the assistance of
the ventilator. The family was told this terrible news, and after a decent
interval, the physicians gently broached with them the issue of whether
to remove the machine and let the woman's body die. At first there was no
response; the question just hung there, deliberately unanswered. But over
the next day or so the physicians persevered. Something, however difficult, had to be done. There was no hope of a change in the patient's condition. Finally, under pressure of the doctors' questioning, the family members raised only a single query: was her heart beating strongly and
independently? Well, yes, the doctors answered; but that was not the
point. It was the injuries to the brain that mattered. "No," said the eldest
son, speaking for the rest, "as long as her heart is beating, my mama has a
chance. When I was a boy, I was brought sick to this very hospital. The

doctors said I would die, but my mama wouldn't give up. She prayed to God that my heart would keep strong, and it did; and here I am today. According to the Bible, the heart is the seat of life. We'll wait and see what happens. We're not going to give up on Mama, any more than she gave up on me. It's all in God's hands."

What were the doctors supposed to do now? From a clinical, and indeed from a legal, point of view, the matter was straightforward. The latest and most reliable diagnostic technologies showed that because of the damage to the brain, she was irremediably and horribly injured. In fact, she was, from the standpoint of brain function, already dead. She would never talk or walk or hope or fear or feel again. It is not the heart but the brain that is the seat of life, or at least the seat of what we would call a human life. It is an error—one might even want to call it a superstitious error—to believe that as long as the heart was beating, there was a chance that this woman would live again. So said the doctors, to themselves first of all, and then to us, the silent onlookers who now judge them.

To repeat: what were the doctors supposed to do? I do not know; but I know what they did do. They took the family at their (reluctant) word and managed this matter as if what stood between the family and a decent disposition of their mother's death and life were a number of false (but comforting) beliefs, beliefs that needed to be corrected in order that justice be done for all concerned. They undertook this correction patiently, sensitively, and with all the goodwill in the world. Their explanations were clear, true, and certainly well meaning. They worked hard to be fair to all concerned. I probably do not need to report the outcome: pain, anger, and a sense of betrayal on the part of the family; prideful pleasure in their self-perceived superiority on the part of the doctors; the threat of lawyers; in short, an ugly muddle. Could things have been different—or better? Would they have been significantly different—or better—if these physicians and nurses had read Wittgenstein on Frazer and had known the story of Mrs. Darlington? Would they have been different—or better—if the woman's family had read and known these things? I cannot answer with any assurance. Thinking about Mrs. Darlington in light of Wittgenstein's remarks about Frazer makes me think differently about that large family waiting patiently in the hospital corridor, about their bedside vigils and the quiet songs they sang to their mother, about their prayers and their (fruitless) confidence in her recovery; but it does not tell me what to do in relation to those things, and I cannot see that it would tell me what to do even if I were a member of the house staff that had—or thought it had—to do something. Faced with rituals and assurances that make no sense to one, what should one do? Confront them? Ignore them? And if one cannot? Mrs. Darlington's grief was, mostly through embarrassment and fear, benignly ignored by her family and friends, and it ran its course with no significant cost to any third parties. The case of this elderly woman was different, or was thought to be. A benign neglect of the situation would have a human and a monetary cost that was thought by those good doctors and nurses involved to be insupportable. Letting things be did not seem to them to be a just and decent option. Scientific ignorance, however comforting to those whose ignorance it was, could not be allowed to control the situation. But was it ignorance (or superstitious error)? The false and naive beliefs the doctors sought to overthrow were only produced under pressure, just as (I imagine) Mrs. Darlington would have produced beliefs about Lee's being alive in heaven and just as Frazer's "savages" would have produced beliefs about better harvests through divine intervention.

What is at issue there in the hospital is not, in the first instance, false belief—though that is certainly the way the matter is most likely to present itself to the parties involved—but a sensibility that cannot help but see actions and reactions as springing from belief, true or false. And both parties to the conflict share that sensibility; both think that they, and their lives, stand firmly on a view of things that is correct. If one were to hope to close the painful gap between this suffering family and the hospital professionals who so sincerely wanted to aid them, it is this sensibility that would need to be altered. Family and professionals would have to come to think of their attitudes—helpless word!—as more original and fundamental than the particular beliefs they precipitate in discussion, and both would have to be able to engage those attitudes as distinctively and fundamentally ethical, that is, as exhibiting forms of character and life both contingent and yet not subject to alteration at will.

Such collisions of ethical sensibility are, for all their commonness, still more likely than not to be intractable, mostly because their character as collisions of sensibility is not easily recognized. And even when it is, we lack, and (so far as I can see and hope) will always lack, any reliable method for preventing or resolving such collisions. It is just here that

Wittgenstein's philosophical thinking can perhaps be useful, since in that thinking, and particularly in the *Philosophical Investigations*, the clash of sensibility (not belief) is made definitive for philosophy itself, and specific suggestions and illustrations are offered in relation to it. For Wittgenstein, the sensibilities that count most in philosophy, in the generation of philosophical perplexity and in the production of philosophical "theories" to clarify what so deeply puzzles us, are rooted in our enchantment by particular grammatical pictures, pictures whose hold on us can only be mitigated by the skillful and timely deployment of other grammatical pictures. The technique of philosophical liberation is "aesthetic"; it focuses on gestalt rather than conviction, or on the principles that generate such conviction. It aims less at changing specific beliefs and more at getting oneself to see something differently, to pick up the stick at the other end.[16]

But even if one is impressed, as I am, with the results such Wittgensteinian operations produce in philosophy, and even if one is persuaded that the same sorts of considerations are in play in the cases I have described, still one may wonder whether the grammatical inquiries of the *Philosophical Investigations* have much of a chance of altering the divergent sensibilities that produce this ethical conflict. It certainly seems clear that no method either could or should be available to settle such quandaries. Any such a method would be horrible, since it would rest on the discovery of a way to guarantee agreement in attitude. Of course, it is true that philosophy has long hoped and believed that such a method could be found, and the construction of ethical conflict as resting on dispute about what beliefs are true (does a beating heart mean that death has not occurred?) has been a key element in the claim that such a method is provided in what we have been taught to call reason or science. But as I read Wittgenstein, he both admires the reach of such construction and laments and fears the sense that it is enough. In some cases of genuine ethical perplexity, perhaps in most of them, the conflict is deeper than some belief that can be settled by "reasonable" appeal to "the facts." Here is the collision of sensibilities, not dispute about *was der Fall ist*.

I have said, honestly, that I do not know what I would have done in that awful situation in the hospital, and that I do not think that my own reflections on Wittgenstein, although they would have made some difference in the way I would have seen the matter, would have made it clearer to me what I should do. But perhaps that is not surprising, or should not be. For whoever thought that philosophy, even good philosophy, would take away the ethical difficulties of life, or even of the practice of medicine? Certainly not Wittgenstein. It would be enough if it could teach us where those difficulties truly lie. Maybe then we could begin to start trying to make a better run at them.

Notes

1 The reference to Mount Nyangani is from Sara Maitland, *Ancestral Truths* (New York: Henry Holt, 1995).

2 That we consider a particular practice superstitious does not mean, of course, that we would be willing to offend its adherents by saying so. It might be decent to play along, consoling the distraught mirror-breaker or refraining from loud noises on the mountain. But that is merely common courtesy, not an acknowledgment that the superstitious practices have some authority for us. It is a nice question when such courtesy becomes a kind of patronizing tolerance that is justly to be criticized.

3 I was once a guest at a student religious retreat in which a Eucharistic celebration was conducted to the music of a jazz ensemble. After the ceremony, I wandered back into the hall, where I noticed the musicians, the leader of whom was the chief celebrant, tossing around the unused loaves of bread as if they were footballs. Though not myself a Christian, I found myself both angry and uncomfortable at the display. Why? Pure superstition? Residual piety? After all, I did not (and do not) believe what Christians believe about the Eucharist.

4 Wittgenstein, PI § 654.

5 Wittgenstein, OC § 559.

6 If we were to come upon a stranger making sounds and acting in ways we could not understand, it might be our best hermeneutic strategy to assume that some (mostly true) beliefs underlie and explain these occurrences, and then to see if we can determine what those beliefs are. I do not deny the utility of the strategy. I do deny that the "picture" (to use a Wittgensteinian word) of human beings that the strategy inclines us to accept (roughly, that all human actions are to be understood as the expressions of beliefs) is the best picture of human beings in all circumstances. As Wittgenstein said, it is not the picture that is the problem for philosophers; it is the application of the picture, and in this case, the application of the "belief" picture to all instances of human life.

7 Ludwig Wittgenstein, "Remarks on Frazer's Golden Bough," trans. John Beversluis, in *Ludwig Wittgenstein: Philosophical Occasions, 1912–1951*, ed. James C. Klagge and Alfred Nordmann (Indianapolis: Hackett, 1993).

8 James G. Frazer, *The New Golden Bough*, ed. Theodor H. Gaster (New York: Criterion Books, 1959), p. 165.

9 One of the remarks that Wittgenstein placed near his remarks on Frazer but that did not make it into his typescript is: "I now believe it would be right to begin my book with remarks about metaphysics as a kind of magic. But in doing this I must not make a case for magic nor may I make fun of it. The depth of magic must be preserved. Indeed, here the elimination of magic has itself the character of magic."

10 William James, "The Will to Believe," in James, *Writings, 1878–1899* (New York: Library of America, 1992), p. 457 and passim.

11 John Dewey, *A Common Faith* (New Haven: Yale University Press, 1934), passim.

12 For the sake of economy, I have ignored desires in my account so far.

13 See Daniel Dennett, *The Intentional Stance* (Cambridge, Mass.: MIT Press, 1987).

14 For example, we might say, "I know what she is visibly doing about Tom, but what I really want to know is what is going on in her heart in relation to him." Here the grammatical picture of inner and outer is perfectly clear, and perfectly useful. The problems arise when we begin to lean on it too hard, expecting it to yield insights into the "reality" we assume to underlie the grammar of our particular idioms. For more discussion of the role of (grammatical) pictures in our thinking, see my *Ethics without Philosophy: Wittgenstein and the Moral Life* (Gainesville: University Presses of Florida, 1982).

15 Wittgenstein, "Notes on Frazer's Golden Bough," pp. 123 f.

16 I have tried to develop these themes at length in *Ethics without Philosophy*.

3 Patient Multiplicity, Medical Rituals, and Good Dying: Some Wittgensteinian Observations
Larry R. Churchill

I'll teach you differences.—Shakespeare, *King Lear* [1]

Man is a ceremonious animal.—Wittgenstein, *Remarks on Frazer's "Golden Bough"* [2]

It would be hard to overestimate the influence of Ludwig Wittgenstein on twentieth-century philosophy. He has few rivals in originality or influence, and because his style of philosophizing is aphoristic, dialectical, and ultimately skeptical, his writing is a continuing source of insight and controversy. Some think this makes Wittgenstein an inappropriate influence on contemporary bioethics; others believe it makes him an ideal one. His influence will be problematic for those who seek a clear model for decisional analysis or an ethical system to extend and apply to the biomedical realm. But he is ideal for those ethicists whose approach is skeptical and anti-metaphysical, yet practical and irenic—that is, for those who believe that the answers are to be found in better questions. To this latter group (in which I like to count myself), hardly anyone can be more important than Wittgenstein, who combines the deep skepticism of Hume, the disdain for convention of Nietzsche, and the sharp observation of Montaigne.

The line from Shakespeare quoted in the first epigraph is one Wittgenstein considered using as a motto for the *Philosophical Investigations*.[3] In his biography, Ray Monk notes that Wittgenstein was deeply impressed by differences of all sorts, and especially by the variety he saw in nature. "The pleasure he derived from walking in the Zoological Gardens had much to do with his admiration for the immense variety of flowers, shrubs and trees and the multiplicity of different birds, reptiles and animals."[4] Wittgenstein's repudiation of the theory-driven formulations and explanations of his own early work stressed differences in the use of language and the pluralism of meaning evident when one stops theorizing and begins to observe. This is typified by his command "don't think, but look!" [5] As Judith Genova aptly puts it: "Concepts are meant to generalize, to bulk experience. Percepts, on the other hand, are meant to particularize." [6] The aim of looking is to notice the differences glossed over in thought.

One aim of this essay is to illustrate the usefulness of the Wittgensteinian imperative "don't think, but look!" for bioethics and, ironically, especially for the more empirically oriented bioethics research now so prevalent. More precisely, I will argue that some of the current research on what is called "patient preferences" for care at the end of life is flawed by assumptions that "bulk" experiences—that fail to notice the multiplicity of values involved as persons seek a death they would call "good." These assumptions lead to dubious conclusions about what sort of medical care is appropriate at life's end.

A second aim of this essay is to call attention to the ceremonious and ritualized aspects of medical practices. There is certainly nothing new in the observation that medical practice is highly ritualized, but its significance is seldom discussed. I will argue that paying attention to the power of medical rituals helps explain why patient "preferences" are often ignored. To say that medical practices are ceremonious and ritualized is to say that they are not undertaken as pragmatic and as means to an end, but are expressions of something more fundamental and unreflective. Rituals cannot be understood as crude and erroneous attempts by "primitive" or superstitious persons to explain things that science explains better, or as vestigial residues of a time when medicine was more magic and religion than science. I will argue, following Wittgenstein, that rituals are ineradicable ways that people participate in and seek harmonious relations with the fundamental realities of their lives.

This theme in Wittgenstein's work emerges most clearly in his *Remarks on Frazer's "Golden Bough."* Wittgenstein was unrelenting in his criticism of Frazer's interpretation of religious rituals as mistaken attempts to predict or control the natural world, that is, as mistaken pseudoscience. "Baptism as washing.—There is a mistake only if magic is presented as science . . . If the adoption of a child is carried out by the mother pulling the child from beneath her clothes, then it is crazy to think that there is an *error* in this and that she believes she has borne the child." [7] Wittgenstein repeatedly criticized Frazer for his provincialism. "What narrowness of spiritual

life we find in Frazer! And as a result: how impossible for him to understand a different way of life from the English one of his time." [8]

The second aim of this essay is to draw attention to the powerful ritual
significance of cardiopulmonary resuscitation. I will argue that one of the
reasons that patient requests to forgo resuscitation near life's end are seldom elicited—and when elicited, often ignored—relates to the power and
importance of resuscitation as a medical ritual. We will be just as provin-
cial as Frazer if we assume that resuscitation is motivated by scientific evidence, or by rational assessments that it routinely achieves the aim of
bringing patients back to life. Why it "works" for health professionals is
more deeply lodged in their sensibilities and therefore more difficult to
change. Recognizing the depth and meaning-bearing capacity of medical
rituals like resuscitation will lead us in a different direction than the one
usually proposed to improve the care of the dying and to satisfy patients'
requests for a "good" dying. While most studies recommend better communication between physicians and patients and better understanding of
the marginal effectiveness or futility of resuscitation, I will suggest that
the chief remedy for unwanted resuscitation attempts is different and better rituals.

The SUPPORT Study

I begin with an examination of the SUPPORT Study, as reported in the *Journal of the American Medical Association*, 22 November 1995, because of its
size and prominence.[9] "SUPPORT" stands for "Study to Understand Prognosis and Preferences for Outcomes and Risks of Treatment." Funded by the
Robert Wood Johnson Foundation, it enrolled over 9,100 patients suffering from life-threatening illnesses in five U.S. teaching hospitals over a
four-year period. Those enrolled in the study suffered from a variety of serious conditions: acute respiratory failure, multiple organ system failure
with sepsis or malignancy, coma, congestive obstructive pulmonary disease, congestive heart failure, cirrhosis, and metastatic colon cancer,
among others. In order to qualify for the study, patients had to have a 50
percent chance of dying over the next six months. The first phase of the
study identified several problems with dying in U.S. hospitals; the second
phase tested interventions designed to alleviate these problems.

The problems identified in the first phase were several. Many patients died after prolonged hospitalization or intensive care, or in pain that was not adequately treated. Roughly half of all do-not-resuscitate (DNR) orders were written just a few days before the patient's death, when a resuscitation attempt would almost certainly have been futile in any case. Ten percent of patients died after spending a month in intensive care. The SUPPORT investigators cited three reasons for these problems: physician uncertainty about prognosis, leading to what the researchers viewed as excessive use of technology; physician ignorance of patient preferences about the use of high tech life-support; and poor communication between physicians and patients or their families. Poor communication and the resulting ignorance about patient desires led, the investigators claimed, to practices in which many patients received life support they did not want.

The interventions of the second phase of the study were designed specifically to alleviate these problems. For example, in the second phase physicians routinely received prognostic estimates about patients and information about patients' preferences. Nurses were trained and employed to act as intermediaries, facilitating discussion among physicians, patients, and families about life-sustaining technologies. The hope was that this intervention would enable patients to have a larger say in their care, with the underlying assumption that if they were given that larger say, they would choose less intensive intervention at the end of life.

To the disappointment of the investigators, none of the interventions seemed to make a difference. Patients in the group that received the interventions still had frequent and substantial pain during their last days and still had DNR orders written only two days prior to death. Fewer than half of the physicians whose patients wished to have life-sustaining measures withheld understood their preference. In short, none of the interventions made a difference in making the care of patents at the end of life less technologically driven, less painful, and more directed by patient wishes to avoid or forgo life support.

While the SUPPORT study has been widely discussed and many of its implications explored, two of its most salient features have not been given sufficient attention.[10] The first is the assumption of the SUPPORT researchers about what constitutes "good" dying, namely, their failure to consider a multiplicity of models for a desirable death. Here Wittgenstein's affinity for noticing and appreciating differences will be salutary. The second problematic feature of SUPPORT is the underestimation of what it takes to change medical practice. Here Wittgenstein's appreciation for ceremonies and rituals will guide my argument.

Patient Multiplicity and Good Dying

The stated objective of SUPPORT was "to improve end-of-life decisionmaking and reduce the frequency of a mechanically supported, painful and prolonged process of dying." [11] The investigators did not seriously consider the possibility that these two objectives might be at odds with each other. In other words, improving end-of-life decision making—by empowering patients to have a larger voice in these decisions—might not reduce the frequency of a mechanically supported, painful, and prolonged dying process, but might rather confirm it or even increase its frequency. (The SUPPORT researchers do acknowledge this possibility in a more recent publication.) [12] It is clear that those patients who wanted less intervention were not heard by their physicians, but neither were those who might have wanted what they received, or even more than they received. I reject the assumption of the investigators that the great majority of patients with poor prognoses will want less aggressive care, or that most patients want a dying process marked by low tech and high touch. I am not, to be sure, claiming that patients routinely seek a painful death or one burdened by useless technological intervention. Yet pain and the temporary use of burdensome technology are precisely the sort of thing many of the severely ill are prepared to tolerate in the hope of recovery, and the usefulness of the technology was precisely one of the things found to be in question in the first phase of the study, expressed as "uncertainty about prognosis." Although the SUPPORT investigators say that the death of many of these patients was "predictable," the interventions of SUPPORT to improve communication may have done little to convince patients (or their physicians) that their illness trajectories were a clear and irreversible one toward death, and that the only question they faced was *how* to die.

This mismatch in the predictability of death between SUPPORT investigators and at least some physicians and patients in the study is surely important. Yet it is the deeper and more pervasive assumption that informed the SUPPORT study that concerns me here: that of the uniformity of patient "preference." This is the assumption that every patient with a poor prognosis, for whom interventions will be at best marginally effective—even patients who know they are dying—will desire the kind of death the SUPPORT researchers find desirable. Other researchers have discovered great variety in what patients with poor prognoses want.[13] For example, Marion Danis and her colleagues have found that the majority of patients they interviewed—patients who, in the degree of critical illness they suffered, were similar to those in the SUPPORT study—"wanted to receive a life-sustaining treatment if it would prolong life for any length of time," and "many patients were willing to have a life-sustaining treatment even if it had no probability of working."[14] In light of these findings, it is at best unclear how often intensive care and other technological interventions ran counter to patient wishes in the SUPPORT study, since physicians who provided these interventions may well have been doing what patients wanted, even if communication was poor. The SUPPORT investigators seem to ignore the normative importance of their own finding that "most patients and families indicated they were satisfied." Moreover, the physicians "felt they were doing the best they could, [and] were satisfied they were doing well."[15]

One conclusion might be that the values of the SUPPORT investigators were truly out of step with those of many patients and their families. Even though many patient deaths were painful, technologically burdensome, and prolonged, and even in cases where communication was poor, in many instances the care given did satisfy all concerned: patients, their families, and their caregivers. Because communication between physicians and patients did not improve in the second phase of SUPPORT, it is difficult to know if this explanation is valid. At a minimum, it is clear that patient desires for a good dying are more diverse and complex than the SUPPORT researchers allowed for. To clarify this point, I offer three groupings for patients, based on their "preferences." This is, to be sure, an oversimplification, but it serves to illustrate that not everyone wants the same interventions in situations of poor prognosis. I call these groups the Fearful Minimalists, the Hopeful Vitalists, and the Anxious Agnostics.

Fearful Minimalists want to exit life with the least amount of expense and fuss possible. Sometimes these persons are motivated by noble aspirations, like saving financial resources for the grandchildren or sparing their spouses and children the pain and disruption of caring for them over a protracted demise. They frequently sign living wills and durable powers of attorney and place them in multiple locations to be sure they are accessible. They show up at bioethics conferences to ask technical questions about the validity of advance directives in various jurisdictions. Not always, but frequently, these persons are motivated by fear. What they fear is what they take to be a fate worse than death—a demeaning, expensive, prolonged dying, with them tethered to some piece of supportive but not restorative machinery, enslaved by an imperative technology, by an institution fearful of litigation or in the grip of a vitalistic, religious dogma, and by a cohort of professional helpers in a passive, self-protective posture, with perhaps a guilt- and denial-ridden family to boot. The Fearful Minimalists see this, rightly I believe, as an almost irresistible set of forces. If they are to have any chance at calling the shots for themselves, they must be well armed with legal documents and willing to risk a premature death in order to avoid a lingering, expensive, painful, or vegetative one.

At the other end of the pendulum swing are the Hopeful Vitalists. They are *vitalists* because they will count any degree or quality of life as worthy of the candle. They see no injustice in spending whatever it takes to preserve their presence. Arguments about the alternative uses of scarce resources in health care bounce right off the Hopeful Vitalists: "To hell with the trade-offs, this is *my life* we're talking about!" Often their vitalism is couched in religious rhetoric about life's "sanctity." They are *hopeful* vitalists because they have bought wholesale the propaganda of medical research and pharmaceutical firms that breakthroughs happen routinely, that a cure for their particular disease is just around the corner, that we are winning the "war" on cancer and heart disease and may soon arrest or reverse the aging process. They are truly hopeful, and not just overly optimistic, because they hold their views not because of but in spite of evidence to the contrary. If one must die, the ethical imperative is not to die too soon. So what some would regard as inappropriately aggressive, painful, and burdensome care, these people want. There are more of us in this category that one might think. The findings of Danis and her co-investigators, as noted above, confirm that a majority of patients want intensive care and maximal treatment to preserve life, even for just a few days. Lots of patients never give up hope of recovery until the very end, if at all. And even if a patient wants to forgo maximal efforts at survival, the family may not. Norman Levinsky voices the experience of many when he says that "it is far more frequent for patients and their families to demand aggressive treatment against the advice of their physicians than for doctors to press to continue therapy that patients or their families want to discontinue." [16] Perhaps in seeking to alter physician behavior, the SUP-PORT study aimed its intervention at the wrong target. In any case, the study seemed not to take seriously the idea that maximal intervention would be what many patients would truly want when they are severely ill.

Occupying the middle ground are the Anxious Agnostics. These persons do not know their "preferences" until they get there, and they are skeptical of both too little care at the end of life and too much. Anxious Agnostics are anxious because they have heard horror stories of both excess and neglect. They are equally anxious about therapeutic zealots, who will continue treatment too long, and efficiency zealots, who, in the employment of entrepreneurial managed care, will stop too soon in order to save money. Anxious Agnostics simply want to die at an appropriate time, without too much expense and too much pain and with minimal family burdens, but they also do not want to go before they really have to, when they will be "ready." And for that signal of readiness they depend fundamentally on their families, their friends, clergy, and especially their physicians. They look to all these persons to confirm their sense that this is a good time to stop, or that it is not. With this agenda, the chief problem unearthed by the SUPPORT study is that patient-doctor conversations about the timing and style of death take place so infrequently.

In brief, one of the chief problems with the SUPPORT study is its simplistic model of the good death, a model that both submerges the potential for variation and is at odds with the findings of other clinical studies about patient "preferences." In this essay, I have consistently placed "preferences" in quotes to note my objection to this term. To say that patients have "preferences" while health professionals gather "evidence," follow "orders," and assume "responsibilities" tends to infantilize patients and trivializes their perspectives. Instead of "preferences," we must begin to call them patient *values*, to signal that patients have depth of conviction and should have weight in the decisional process. I have preferences about the color of my tie, but I have deeply held values about how I should die and how much effort should be expended on my care in extremis. Care of the severely ill and dying cannot be improved so long as bioethics researchers continue in this paternalistic tradition of referring to patients.

More generally, bioethics researchers would do well to let patients and their families teach us about the variety of values that inform good dying. Attention to these differences is an elementary requirement for valid research, as well as for patient-centered care. As Arthur Frank puts it so aptly in *At the Will of the Body*, "Care begins when difference is recognized." [17]

Medical Rituals and Good Dying: Why Change Is Not a Matter of Communication

Early in this essay I defined rituals as ways that persons orient themselves to what is important in their lives, ordering their experiences to seek harmonious relations with their world. This definition is evident in the etymology of the term. *Webster's Ninth New Collegiate Dictionary* defines *ritual* as "the established form of a ceremony" and cites its Old English source as *rim*, meaning number. Rituals give to ceremonial events an order to follow, as if numbered. This ordering is especially useful when the life events being interpreted are stressful, involve passages into the unknown, or are marked by uncertainty or danger. Resuscitation is clearly such an event and can be fruitfully explored as a ritual of medical and hospital practice. This exploration will deemphasize the issue of whether it is technically useful and focus on what function it serves for physicians and hospital staff. Whether it "works" is therefore a question of whether it successfully bears the weight of meaning given to it and orients those who are engaged in it, not whether the patient is revived. Approaching resuscitation in this way will help us understand one of the reasons why relatively few DNR orders are written until very late in the course of a dying process and why patient requests for DNR are often ignored, as SUPPORT and other studies have shown.

Rituals promote solidarity. Resuscitation requires a team, working rapidly and in concert, at the edges of life. If health professionals go through resuscitation together they are likely to bond, to feel closer, and to confirm for each other the validity of the experience. This occurs whether the attempt succeeds or fails. The psychology of group cathartic experiences is at play here, and it is available not only to health professionals, but to all of us, at least vicariously. Weekly doses of *ER*, Chicago Hope, and *Rescue 911* make us all witnesses and empathic participants in resuscitative efforts, which in the television versions brings the patient back a remarkable 75 percent of the time![18]

Rituals enhance communication. They provide a normative way for talking about these stressful, life-changing events. They do not, however, provide a language of critical interpretation. Rather, the language around rituals tends to confirm the ritual values, to validate and reinforce the meaning provided in the ritual. Thus, the idioms available through the ritual language do not concern whether a resuscitation should have been attempted, but the technical reasons it succeeded or failed. Was the response time adequate? Was the crash cart properly supplied? Were the right techniques used? Rather than addressing the grounds for attempting or forgoing resuscitation attempts, these idioms supply language for an exchange between insiders. This language, then, serves as a point for professional reinforcement about the central value that resuscitation efforts embody, namely, the preservation of life. Resuscitation attempts, regardless of their success, remind health professionals that they are committed to life, a value dramatized for public endorsement through weekly television.

Rituals give confidence in the face of danger. And dying is very dangerous, at least to the modern American sensibility. Resuscitation gives us something to do, so we will not be left just standing there, without an active role to play. By having a prescribed role, health professionals gain confidence. The opposite of having confidence in our roles in the face of death is likely to be panic and guilt. Resuscitation attempts are the medical and hospital equivalent of Catholicism's last rites. The sin is to die without receiving them.

Rituals provide those who participate in them with power. This is a summation of the three features mentioned above—solidarity, language and confidence—but it goes beyond them. Resuscitation attempts—again, just the attempts, never mind their success—give those who participate in them power over death. This is not primarily a power of the mind, but a power in the hands, felt and exercised in the muscles. To understand the ritual power of resuscitation is to recognize that entering the body of another with needles and tubes and pounding his chest with fists or electric shock can energize and reassure us about our own mortality. Physical exertion of many sorts is a way of fighting for our own lives; for example, many of us exercise to escape the demon of our mortality, both literally and figuratively. As James Dickey wrote regarding the exercise he did to manage his diabetes:

> Each time the barbell
> Rose   each time a foot fell
> Jogging, it counted itself
> One death   two death   three death   and   resurrection
> For a little while.[19]

Resuscitation, when it succeeds, shows that we have conquered death, at least "for a little while." When it fails, it provides the reassuring demonstration that although our patient is dead, we are alive.

Finally, rituals are powerful and important because of the deep connection between "rite" and "right," between what is ritually appropriate and what is morally good. This explains why rituals persist even when we have abandoned the belief system that explains their larger purpose. People pray long after they have ceased to believe that there is a god on the other end listening. They pray because the ritual of prayer is itself comforting, even if sheared of its metaphysical accompaniments; it seems like the "right" thing to do, or sometimes the only thing we know to do. Like prayer, resuscitation can serve in a crisis as the default position. Also like prayer, resuscitation is more deeply rooted in our bodies and our emotions than in our critical intellect. We should not be surprised that it is hard to change.

Acknowledging all these powerful features at work, we can see why the SUPPORT study's well-intended and straightforward efforts to improve communication between doctors and patients made no difference. Communication problems are, no doubt, real issues of concern, but lack of communication was not the chief reason why DNR orders were written so near death, nor why they were ignored. Better care of the dying means developing and enacting the rituals of palliation and hospice care. Absent this, improved communication is likely to fall on deaf ears. Patient requests to forgo intensive efforts at life support in hospital settings invoke rituals health professionals have yet to master, in a language they do not yet fully understand.

John Lantos, describing the resuscitation efforts he and his wife enacted at the death of his sister-in-law and the subsequent gratitude of his relatives, concluded: "Until we come up with alternative rituals, other ways of dramatically affirming and valuing the lives of persons who are sick and dying, we will need these rituals badly; even if (or perhaps especially when) they are mostly symbolic." [20] My claim in this paper echoes that of Lantos and tries to specify more concretely what those alternative rituals might be.

Better care of the dying depends at least in part on the acceptance of palliation and hospice care as medical and hospital rituals. This means more than a theoretical acknowledgment that hospice care is available and a possibility for patients with severe, irreversible illnesses. For practices to become ritualized, they must become part of the customary way of doing business, one of the first things that come to mind, rather than an alternative that is considered after all else fails. Part of the problem is ignorance: many health professionals have only a superficial understanding of hospice care rather than a working, on-the-ground knowledge of it. Part of the problem is bias: many health professionals are uncomfortable with hospice because they associate it with attitudes of defeatism, and they see themselves as being obligated to "err on the side of life." Medicine is perhaps more widely infected with this belief than nursing and the other health professions. Still another part of the problem is incompetence, insofar as few health professionals are really skilled at palliation. Palliative care is not a routine part of medical education, and initiates have little chance to become practiced and competent. In summary, ritualization means that a practice is routine in the habits of health professionals, is part of their list of competencies, and has a meaningful place in their collective psyche, that is, in the overall ethos of patient care plans. Palliation and hospice care do not yet have that status. Hopefully they will in the future.

I have been critical of the SUPPORT study in its assumptions and in its overly intellectual efforts at change. Yet I share many of their assessments about what would count as a good death. My argument has not been that the image of death and dying they advocate is not a good one for us both as individuals and as members of compassionate communities and a just society. Rather, my point is that there is little evidence that this image is shared by the majority of patients when they are critically ill with poor prognoses, or by the health professionals caring for them. Perhaps the SUP-PORT study is ahead of its time. I genuinely hope that the values of its researchers bespeak a place to which we are moving in the care of the severely ill and dying. Getting there will mean paying far more attention to differences, and remembering, as Wittgenstein noted, that we are ceremonious animals.

Postscript: A Wittgensteinian Admonition

In the beginning of this essay I emphasized some motifs of Wittgenstein's later work that can be useful for assessing empirical bioethical research, such as attention to detail and differences, maxims about looking as well as thinking when doing research, and reminders that the researchers should attend to rituals and may themselves be captive to rituals of both thought and practice. Those familiar with Wittgenstein's work will be readily reminded of a frequently quoted phrase from the *Philosophical Investigations* that can serve as a way of framing the hazards of bioethical work more generally, and it is with this admonition that I close. I refer here to Wittgenstein's confessional statement about what was amiss in his early philosophy, a statement that echoes his disdain for generality and abstraction.

In speaking of his early work in the *Tractatus logico-philosophicus*, Wittgenstein said, "One thinks one is tracing the outline of the thing's nature over and over again, and one is merely tracing round the frame through which we look at it." [21] He went on to observe, "A *picture* held us captive. And we could not get outside of it, for it lay in our language and language seemed to repeat it to us inexorably." [22] Wittgenstein was speaking here of his own early efforts to make language picture the world, that is, to make the logical form of propositional assertions the sole way to represent reality.

The more general implication for bioethics is apparent. It is not just empirical work in bioethics that gets captured by our pictorial biases. Bioethical theorizing is also subject to this pitfall. Not just Kantian or utilitarian requirements for ethics, but also principlist, narrative, feminist, or postmodern pictures, can obfuscate ethical seeing and thinking, can hold us captive and repeat the "facts" of an ethical problem over and over to us in a singular and privileged idiom, as if they *must* be true. The remedy is a healthy skepticism about our ability to theorize reliably without remainder, and without considerable self-deception, and a commitment to reframing, not simply retracing the same outlines of thought. Wittgenstein suggested that philosophers need to bring words back from their metaphysical to their everyday use. Bioethicsts need to bring ethical theories back from their metaethical to their practical uses.

Notes

I thank Marion Davis and Carl Elliott for their careful reading and helpful comments, and Roy Martin for his encouragement and invitation to try out the ideas in this essay at a Community Ethics Conference in Fort Worth, Texas, in 1997.

1 Shakespeare, *King Lear*, act 1, scene 4.
2 Wittgenstein, *Remarks on Frazer's "Golden Bough,"* rev. and ed. Rush Rhees, trans. A. C. Miles (Atlantic Heights, N.J.: Humanities Press, 1979).
3 Ray Monk, *Ludwig Wittgenstein: The Duty of Genius* (New York: Penguin, 1991), p. 537.
4 Ibid.
5 Wittgenstein, PI § 66.
6 Judith Genova, *Wittgenstein: A Way of Seeing* (New York: Routledge, 1995), p. 57.
7 Wittgenstein, *Remarks on Frazer's "Golden Bough,"* p. 4.
8 Ibid., p. 5.
9 The SUPPORT Principal Investigators, "A Controlled Trial to Improve Care for Seriously Ill Hospitalized Patients," *Journal of the American Medical Association* 274 (1995): 1591–98.
10 See, e.g., the ten short essays assessing the SUPPORT study in the *Supplement to the Hastings Center Report* 25 (1995): S1–S36; see also the "Sounding Board" essays by Norman Levinsky and Daniel Callahan in the *New England Journal of Medicine* 335 (1996): 741–46 and Bernard Lo's editorial accompanying the SUPPORT report, *Journal of the American Medical Association* 274 (1995): 1634–36.
11 The SUPPORT Principal Investigators, "A Controlled Trial," p. 1591.
12 Joanne Lynn et al., "Perceptions by Family Members of the Dying Experiences of Older and Seriously Ill Patients," *Annals of Internal Medicine* 126, no. 2 (1997): 97–106.
13 See, e.g., L. L. Emanuel et al., "Advance Directives for Medical Care: A Case for Greater

Use," *New England Journal of Medicine* 324 (1991): 889–95; and M. Danis et al., "Comparison of Patient and Physician Choices for Life-Sustaining Treatments," *Journal of General Internal Medicine* 10 (1995): 107.

14 Marian Danis et al., "A Prospective Study of the Impact of Patient Preferences on Life-Sustaining Treatment and Hospital Costs," *Critical Care Medicine* 24 (1996): 1811–17.

15 The SUPPORT Principal Investigators, "A Controlled Trial," p. 1596.

16 Norman Levinsky, "The Purpose of Advance Medical Planning. Autonomy for Patients or Limitation of Care?" *New England Journal of Medicine* 335 (1996): 742.

17 Arthur Frank, *At the Will of the Body* (Boston: Houghton Mifflin, 1991), p. 45.

18 S. J. Diem, J. D. Lantos, and J. A. Tulsky, "Cardioresuscitation on Television," *New England Journal of Medicine* 334 (1996): 1578–82.

19 James Dickey, "Diabetes," in *The Eye-Beaters: Blood, Victory, Madness, Buckhead, and Mercy* (New York: Doubleday, 1968), p. 7.

20 J. D. Lantos, "Bethann's Death," *Hastings Center Report* 25, no. 2 (1995): 22–23.

21 PI § 114.

22 Ibid., § 115.

4 "Unlike Calculating Rules"? Clinical Judgment, Formalized Decision Making, and Wittgenstein
James Lindemann Nelson

In an academic medical center in the southeastern United States, week-days at 7 A.M. means morning report. A group of clinicians—an academic faculty physician, a community-based doctor, six or seven residents, maybe a couple of medical students, perhaps a social worker, a nutrition-ist, a pharmacist, and even a drowsy medical ethicist—comes together to discuss the patients assigned to their service.

These people gather for two reasons, chiefly: to arrive at decisions con-cerning patient care and to teach how such decisions are made to the clinicians-in-training around the table. After roughly an hour of this kind of discussion, the team breaks into two groups, spending the next hour or so visiting patients and continuing to discuss the decisions, sometimes with their patients and families, invariably among themselves.

Such scenarios are played out in thousands of medical centers every morning, no doubt, and have been for a good many years; it is a kind of proceeding that has a venerable legacy. As a philosopher who has been privileged to observe it on many occasions, I am often reminded of Aris-totle's notion of how one achieves competence as a moral agent, namely, by scrutinizing those who have already attained such status as they en-gage in deliberation, decision, and action, trying to figure out what they do, and doing likewise.

Aristotle's idea was that the competent maker of moral decisions pos-sesses a faculty he called *phronesis*, typically translated as "practical wis-dom." Phronesis involves the ability to discern the best course of action in a given instance of choice. It is decidedly not a mastery of algorithmic decision procedures, although of course a person of practical wisdom will have to know a great many general things, both about the world and about its value. Among these will be certain rules and principles and how to draw valid inferences. But seeing where right action lies between vicious extremes is not a matter of deducing conclusions from premises; it is, as we say, a matter of judgment.

The significance of judgment can be detached from Aristotle's general notion of ethics. Judgment is a crucial part of moral deliberation and action on almost any account, for people often encounter situations where what Martha Nussbaum has called the "standing moral terms" underdetermine what ought to be done. Many of us may accept that something like a general requirement of respect for persons, or of enhancing welfare, or of paying special attention to our intimates, is pertinent to a given choice; we accept that rules against violence and lies are relevant to how we think about how we should live. But in the complexities that characterize people's lives, such ideas cannot be relied upon in any clear way uniquely to determine what constitutes admirable action and admirable lives. What it is to express our respect, our care, our love for *just* these people, in *just* these circumstances, in the light of a particular history and a particular set of anticipations, with these uncertainties and these sources of clarity—solving this problem requires judgment.

But the underdetermination of decision by rules and principles is not a circumstance unique to moral decision making. Playing chess, choreographing a dance, choosing to whom to make love or a loan, shifting into third gear, arranging flowers, deciding when to shoot and when to pass, dropping your "clingy" child off at day care—all these things may be done more or less well, and excellence in them is achieved not by memorizing formulae but by developing a highly nuanced sense of what is fitting given the particularities of the presenting situation. So, too, for what goes on at morning report: arriving at a diagnosis, projecting a prognosis, and recommending a therapy. General information is pertinent but not always determinative; grasping the significance of particular features of this patient and her illness is required if excellence is to be achieved.

Or at least, such has been the traditional view. The ritual of morning report, then, is in important part an effort to take people who already grasp a great number of "rules and principles," a lot of pathophysiology and anatomy, and turn them into people who can also see how such gen-

eral notions bear on the highly particular individuals who present themselves for the care of these clinicians.

But where there are rituals, so also are there iconoclasts. Very powerful
reasons exist for calling into question the idea that seems presupposed by
the morning report approach of what it is to become a good clinical decision maker, and there are those who are not at all shy about pushing those
reasons. Recent years have witnessed a struggle over what clinical decisions are and how they should be made—that is, a struggle over the epistemology of the clinic. In this essay, I aim to enlist some insights suggested
by readings of Wittgenstein, not so much to resolve, but to help redirect,
this struggle.

Two Models of Decision Making

Despite its ubiquity in practice, theorists have a long history of restiveness
with the reliance on judgment; it seems, after all, so vague, so inexplica-
ble, so idiosyncratic. Ethics is hardly exempt from this urge: utilitarianism, for instance, might be seen as the attempt par excellence to algo-
rithmatize morality. If a domain that seems so clearly as morality to lend
itself to judgment can have so powerful and persistent a theoretical recon-
struction as utilitarianism, it is not at all puzzling that in areas where discerning what counts as a good outcome is less contestable—bridges either
stand or fall; patients die or they do not—reliance on judgment is highly
suspect.[1]

In my catalog of practices of judgment, I mentioned "deciding to whom
to give a loan," something that may occur from time to time in many people's lives. There are, of course, people who make such decisions for a living, and there was a time when experienced financial officers made judgments about who did and who did not qualify in a way not too different
in kind from how amateurs do it. While there are surely things a prudent
lender needs to know about resources and history, there is also the need to
"size up" a person, to weigh her character, and experience can lead to
skilled judgment about such matters. According to Michael Scriven, there
was a time when such experts could beat the best available "rules"—formulae that correlated histories of repayment behavior with financial and
other circumstances of those asking for the loans—in making predictions

about who would default and who would repay. But the formulae got better, and eventually they were able to outperform the experts. Besides, if the experts were making decisions about who gets what in ways that were not wholly explicit, all kinds of nasty factors, quite irrelevant to the likelihood of redeeming loans, might be influencing their decision making. Much fairer, as well as more accurate, it would seem, to downplay judgment and rely on clear, public, and explicit formulae.[2]

Similar problems seem to infect clinical reasoning and decision making. Scriven claims that, just as there are available formulae that outperform bankers' expert judgment in giving out loans, so too there are relatively simple formulae that outperform physicians' clinical judgment in diagnosis, and we could generate more of them if we were serious about the matter. In the 1979 essay on which I draw, Scriven writes:

> The two most interesting facts about clinical judgment are that it does sometimes turn out to be right even when the judge cannot explicate his or her reasons for the judgment; and that, on the other hand, it is virtually always *less* reliable than a simple formula relating *some* relevant antecedent variable to the criterion variable, *when* that formula has been empirically fitted to previous data.[3]

So we encounter the very disturbing suggestion, based on decades-old work on clinical decision making, that maybe the Aristotelian approach to clinical decision making, and therefore to clinical education, is wrongheaded, or at least obsolescent. And more recent developments make matters look still worse for clinical judgment: studies of clinical practice have revealed disturbing "small-area variations" in therapies provided for patients presenting with the same indications. What will get you drugs in Knoxville, Tennessee, might get you surgery just across the state line in Asheville, North Carolina, despite the fact that there seems to be no reason to suspect that there is anything about the patients living in these two areas that explains the difference in clinically relevant terms.[4] Probably the leading student of small-area variation, J. E. Wennberg, attributes these differences to a physician's personal style, to regional fashion, to convenience, to finances—in short, to factors that seem to be completely detached from why people are ill and what might make them better.[5]

In part, then, as a general suspicion about "judgment," in part as a reaction to what seems distressingly arbitrary in physician decision making and behavior in particular, there has been a powerful move toward replacing reliance on the diagnostic and therapeutic discretion of physicians with reliance on more objective methods that can yield more algorithmic approaches to patient care. "Technology assessment" research and studies statistically assessing the "outcomes" of different therapies propel the development of the algorithms, the clinical "guidelines" or "pathways," that are to direct what doctors do. Many see this development as the key not only to improving patient care, but also to identifying and eliminating waste as the guidelines are progressively incorporated into the structure of managed care.

Some physicians resist this trend, seeing it as a threat to their professional autonomy, as reducing them to practitioners of "cookbook medicine."[6] But while concerns of this sort are understandable from the practitioner's point of view, they may seem relatively unimportant from the perspective of patients and payers; job satisfaction is not an ethically irrelevant matter, but if better and cheaper medicine comes from cookbooks, so be it. Yet physician resistance to what is seen as overreliance on algorithm has also been allied to the idea that the technology assessment–outcomes studies–clinical pathway nexus will actually end up harming patients. Physicians are sometimes dubious about the reliability of outcomes research funded by parties with a stake in how the studies turn out. They also often point out that they do not treat "standardized" patients of the sort making up populations in outcomes research. Not infrequently, some doctors claim, the patients with whom they work will present variations in the course of their disease, their general history, or their response to therapeutic agents that will be clinically relevant. As such physicians see it, a good doctor must have the ability—and the freedom—to be on the lookout for the relevant particularities of each clinical encounter and respond accordingly.[7]

From this debate, I want to extract two models of what good clinical decision making involves. These models are not entirely unrelated, but they are distinct enough that they suggest different forms of research into clinical effectiveness, as well as different forms of medical pedagogy and practice.

I will call the approach that relies on outcomes research and generates clinical guidelines the *formal evidence model* and refer to its champions as *formalists*. If pressed, formalists may admit that rules derived from such research may not be able to uniquely determine every clinical decision. But such rules certainly could do so much more often than is in fact now the case. The extent to which evidence-based rules cannot determine diagnosis and therapy is not to be celebrated as an opportunity for brilliant human intuitions. Rather, it is to be regretted, since it is an apparently unclosable portal through which arbitrariness can pass unchecked into deliberation and decision.

Good clinical practice should strive to squeeze that portal shut as much as possible, and to heavily screen whatever openings are left. There are objective, scientifically justifiable truths that represent an important and distinct part of clinical decision situations, and there are ways and means of improving physician access to the clinical implications of these truths: algorithms, explicit decision procedures, clinical pathways, and computerized expert systems. In many cases, better information of this kind, targeted in the right way, might be able to greatly reduce or even, for all practical purposes, eliminate uncertainty. In cases where the information cannot determine what counts as optimal care—because, say, of very idiosyncratic features of a given case—at least it may be able to inform and limit our sense of the viable clinical options. Being a competent clinical decision maker is crucially a matter of knowing which algorithms for which indications are currently the best supported by controlled studies and other forms of research.

The alternative position I will refer to as the *expert judgment model*. It sees making good clinical decisions as involving a kind of integration of scientific information and scientific models with clinical experience and, perhaps more broadly, cultural understandings and life experiences. This conglomeration gives rise to an individual's ability to see clinically relevant particulars in a way enlivened by their own interplay, and their interplay with an unruly mixture of general ideas and experiences that can suggest fruitful analogies and illuminate the contours of useful patterns. On such a basis, the woman or man of good clinical judgment appreciates the clinical situation in a fashion that is not in principle fully explainable in terms of the explicit rules and codifiable information that could be contained in clinical pathways or expert systems. Technology assessment and outcomes research should inform, but cannot exhaust, how doctors decide. Having good clinical judgment, so understood, has been likened to having a good sense of humor, or a keen sense of position in chess.[8]

Now, it might be thought that any contest of these models ought to be resolved strictly by experiment. It should be, after all, an observable matter whether one or the other (or some third alternative) more reliably leads physicians to beliefs and actions that are in the best interests of their patients. Rather than think any philosopher's views about knowledge, judgment, or decision making should tip the scale for one view rather than the other, philosophers should affect a proper spirit of naturalistic humility; the outcome of the empirically vetted contest between the models should inform epistemologists' reflections about such matters, not the other way round.

While there is surely a great deal to this point, it misconstrues the character of the contest in which these models are now engaged. In the current climate of opinion, some kind of philosophical motivation of the judgment model seems required, simply to show that it is a starter. It now appears far from likely that the strongest versions of these models will be fairly tested against each other. For a number of reasons—some more political and economic than epistemic or ethical—the winds seem to be blowing strongly in favor of the formal method. There is a great deal of money being spent on assessment, measurement, and the development of clinical guidelines and pathways. A major funding source for such work has in the recent past been the U.S. Agency for Health Care Policy and Research, but pharmaceutical companies, eager to demonstrate the superiority of their own drugs, are also in the business of bankrolling outcomes studies. The motivation is in part concern about small-area variations and the pervasive sense that medicine's armamentarium is full of inadequately tested therapies. But this bias is also based, I venture to suppose, on a rather Whiggish enthusiasm for the method of science and suspicion of the unformalized, garlanded with hopes that there is money to be saved (or made) in "normalizing medicine." Or so I interpret the fact that the superiority of formal to more intuitive forms of clinical decision making seems presupposed by research strategies and funding priorities now in place, rather than considered as a thesis to itself be tested empirically. There seems to be little enthusiasm on the part of the most powerful supporters of research on clinical decision making for performing meta-outcomes studies, comparing judgment and guidelines.

I do not mean to suggest that lack of motivation is the only problem with head-to-head testing of the models. What would such a test look like? Presumably, physicians would be randomized to a "judgment" group and a formal decision-making group. But what would be the constraints on each group? Would doctors in the judgment arm of the trial be barred from access to the data and guidelines available to physicians in the formal evidence arm, or would they be allowed to use it as they saw fit? Would those in the formal evidence arm be obliged to treat patients in ways that deviate from their own best judgment of what would be in the patient's interest? It might be tricky, to say no more, to find physicians who would be able to take a stance of "clinical equipoise," not with respect to a particular intervention, but with respect to their own global judgment as physicians. Achieving decent informed consent from patients might be difficult too ("Some physicians believe that patients receive the best treatment if their care is strictly in accord with the results of large-scale research, even if their own personal judgment would disagree with the implications of those results for a given patient under their care. Other physicians think the best results are achieved if doctors have the freedom to ignore the best scientific evidence and treat according to their own best judgment. If you agree to participate in this study, you will be randomly assigned . . .").

Such problems of the ethics of research design are by no means necessarily insurmountable. But solving them is not likely to absorb much attention so long as there remains a powerful institutionalized momentum in favor of increasing the formalization of medical decision making without comparing it carefully to the judgment models. For example, prevailing opinion about these matters would seem to exclude exploring the possibility suggested by Scriven: although many clinicians may exercise their judgment in ways that do not necessarily outperform what formalized methods achieve in the aggregate, some clinicians—diagnostic and therapeutic "superstars"—relying on more-intuitive forms of clinical judgment may indeed outperform the formal models. Mounting a study of Scriven's suggestion would not seem to require a randomized controlled test. But it would require acknowledgment on the part of partisans of formalization that there was something to the notion of clinical judgment in the first place. If we do not regard judgment of this kind as a model worthy of exploration and not merely of elimination, we will not identify these people and learn more about why it is they do what they do so well.[9]

Wittgenstein and Judgment

It is at this point that some remarks of Wittgenstein's look intriguing, not because they can resolve the empirical issue between the proponents of judgment and their formalist opponents, but because they might help keep alive the question of which model, or which relationship between the models, is best. I look at three themes in Wittgenstein's oeuvre—his remarks about science; about interpretation, rule following, and practice; and finally, about expert judgment—in the expectation that they might dissolve some confusions and allow us to see more perspicuously what is at issue between the models. The notion that "the methods of science" represent some sort of general epistemic pinnacle is a topic to which Wittgenstein had something to contribute. Further, unbridled endorsement of formalizing every possible bit of decision making seems to aspire to the goal of making clinical encounters fully governed by explicit rules. But Wittgenstein also had a good deal to say about how rules are indefinitely interpretable and how their final use is grounded in what he referred to as "practices," founded in forms of training that install in language users knowledge of "how to go on" despite this in-principle-indefinite ambiguity. Finally, Wittgenstein also drops a suggestive remark about expert judgment itself and how it is different from skill at using "calculating rules."

The goal of this Wittgensteinian excursus, then, is to show that there may be some philosophical motivation for taking the expert judgment model seriously, and that we therefore have reason to keep open an epistemic option that current work on clinical decision making, as well as current financing pressures in health care, seems bent on closing. But the enterprise is a delicate one. The considerations that lead to what in my judgment constitutes a neglect of good epistemic practice in assessing the strengths and weaknesses of these models are varied, and it is likely that no one response will clarify the matter for everyone who is confused about it. Nor do the comments I assemble constitute knock-down, drag-out arguments in favor of a particular way of seeing the models, or the appropriate ways they should be tested. In fact, part of what I want to do here is to show the limits of appeal to Wittgensteinian inspirations. What I expect is that insofar as the rather narrow way in which the contest of these models is now seen rests on certain sorts of philosophical notions, a good dose of Wittgenstein might well incline people to take a more balanced view of the matter. This may not seem a fully satisfactory way of proceeding; it does, however, strike me as identifiably Wittgensteinian in spirit.

Suspicions about Enthusiasm Concerning Science

I have already suggested that a less-than-critical enthusiasm about science might be "the decisive move in the conjuring trick" (PI § 308)—the point at which the epistemic battle between formalization and judgment is decided before it can really start. Wittgenstein was dubious about this enthusiasm. In the *Blue Book*, he writes, "Philosophers constantly see the method of science before their eyes, and are irresistibly tempted to ask and answer questions in the way science does" (BB, p. 18).

Why is this temptation so irresistible? Presumably because science is our great epistemic success story; unlike other heuristic narratives, and despite its own persistent internecine quarrels, science compels convergence, respects no cultural barriers, and carries all before it. If one took the job of philosophy to be like that of science—to be a "physics of the abstract," as Wittgenstein disdainfully characterized C. D. Broad's conception of the subject—then the enthusiasm of philosophers for science is nothing to wonder at.

Why is it unfortunate for philosophy to succumb to this temptation? Because, as Wittgenstein sees it, the job of philosophy is not like that of science. Roughly put, the point of philosophy is not to add to our knowledge of the world, but to resolve confusions caused by our misunderstandings of the way we represent the world in language. And it does so not by constructing theories, but by paying attention to what is in fact done by human beings in particular settings using language in particular and highly varied ways.

But even if beguilement by scientific visions may be a problem for the way philosophers ask and answer questions, why should we think it is a problem for physicians? After all, physicians do train in the life sciences, and many of their most spectacular results, and the concomitant cultural authority physicians enjoy, are most plausibly seen as flowing from the relationship between medicine and science. Why should the inaptness of science as a method for philosophers entail that it is inapt for doctors? Is there any confusion inherent in thinking of medicine as a "biology of the particular"?

The answer to this question lies, I think, in just how sensitive to particulars clinicians will be prepared to be by formalized regimens of training and practice. If physicians are trained to practice a medicine "of the aggregate," there is a danger that they will lose the habits of perception that allow them to see the significance of how a disease manifests itself in the bodies—and in the lives—of particular people.

Wittgenstein's distrust of science's wiles seems closely allied to his worries about the philosopher's "craving for generality" (BB, p. 17) and her "contemptuous attitude toward the particular case" (BB, p. 18). Infatuation with the general seems part of "what opposes . . . an examination of details in philosophy" (PI § 52). These are all prime pitfalls for Wittgenstein's conception of philosophy. And despite the fact that medicine has a closer connection to science than does philosophy, a regimen of training and practice that encourages any kind of cavalier spirit toward the specifics of the case at hand would seem to be bad news for physicians, and for their patients, too.

However, while a lack of attention to particulars is surely problematic for physicians as well as for philosophers, there are at least a couple of reasons to think that we have here nothing better than a reason to be careful about how guidelines and pathways are received by the profession and imparted to practitioners. For one thing, not all scientists are so beguiled by the image of science as the great abstract totalizer: think of the patience and sensitivity to detail that archaeology demands, for instance. For another, the wisdom of succumbing to temptation is always, in part at least, a matter of what one's alternatives are. Whether there are good alternatives to a progressive formalization of medicine is just what is at issue here. It is certainly plausible to think that physicians also have reason to be wary of contempt for particular cases and the need to examine details, and reasonable to fear that too much pedagogic attention to developing and following guidelines may blunt the sensitivities that orient clinicians to particulars. But whether as a whole more patients would suffer from this kind of perceptual coarsening on the part of physicians than would bene- fit from the removal of unmotivated variations and uncertified practices is surely a question that is both pertinent and fundamentally empirical. Medicine is not philosophy.

So, "nothing better" than a reason to be careful. But also no worse. To the extent that excitement over formal methods rests on the notion that clinical guidelines based on statistics generated from large, controlled outcomes studies simply must be better than clinical judgment because, after all, they are more scientific, Wittgenstein's reminders about the im- portance of particulars support looking more carefully at how clinicians actually make their judgments, and also searching more carefully for ways to compare various styles of clinical decision making. It might seem odd to be enlisting Wittgensteinian caveats about science in support of the claim that we need to be more empirical about the best relationship be- tween formal methods and clinical judgment, but I think this conclusion has no more than the air of paradox. Wittgenstein's warning is not against science, but against a habit of thought, against the sense that any form of reasoning that is scientific must in virtue of that fact alone be superior to any reasoning that operates differently. If we are freed of that confusion, we can then try to get a better hold on the relative bona fides of con- tending approaches, an effort in which "denk nicht, sondern schau!" (PI § 66) may be excellent advice.

I now turn to Wittgenstein's remarks on rules. Here, the support for the suggestion that responsibly preferring formal methods to clinical judg- ment requires at the least a fair trial of their respective strengths is less straightforward, owing in part to the extreme generality of the scope of Wittgenstein's ideas. But his remarks will—or so I argue—prepare us to appreciate a complexity in how decisions are made in everyday contexts.

Rules, Practices, and Judgment

I opened this essay by invoking morning report: a moment in medical ed- ucation, a frozen frame in the extended process of inculcating initiates into a certain organized, regular, and reiterated form of social activity. Wittgenstein referred to such activities as "practices," and he had some rather intriguing things to say about them: "Not only rules, but also examples are needed for establishing a practice. Our rules leave loop-holes open, and the practice has to speak for itself" (oc, § 139). This might seem to provide some aid and comfort to the judgment model, but it is a rather dark observation. Two points need to be clarified: how it is that rules leave loopholes, and how examples fill them in—how, that is, they enable the practice to "speak for itself."

That rules are porous is not an insight particular to Wittgenstein. For example, in the *First Critique*, Kant points out that the process of applying concepts to objects cannot be exhaustively determined by rules, since, were this the case, we would be caught in a vicious infinite regress: each rule would demand a rule for its proper application, which rules would in turn require further rules, and so on, ad infinimitum. Kant ends this regression in a faculty he refers to as *Mutterwitz*, or "mother-wit," which— interestingly enough—is typically translated as "judgment." [10]

But Wittgenstein was deeply interested in the porousness of rules; a good deal of his later thought is occupied in exploring this idea. In addition to the remark from *On Certainty*, consider, for example, a well-known passage from the *Philosophical Investigations*: "This was our paradox: no course of action could be determined by a rule, because every course of action can be made out to accord with a rule" (pi § 201). Here, the Kantian version of rule indeterminacy is given a bit of a twist—the idea is not simply that every rule would need a rule to explain how it works, but that any rule can itself be interpreted in indefinitely many ways. But the problem is the same. How do we nip in the bud the apparently inexhaustible sequence of rules and their possible interpretations? Wittgenstein grounds our competence in applying concepts in "practices": mastering a language, and hence knowing how to wield a great many concepts, requires that a community of some sort initiate the learner into the various ways of seeing and of going on to which that group regularly resorts. The in-principle-uneliminable ambiguity of rules is resolved in practice by having certain patterns of response installed and maintained in us through various forms of social training.

But Wittgenstein's route from this observation about the underdetermination of rules to judgment may be less direct than Kant's. Judgment, for Kant, seems a kind of discrete thing, a faculty inherent in persons whose nature we might come to know better. However, Wittgenstein looks not to something that we as individuals possess, but rather to patterns of behavior installed and reinforced socially, to practices that constitute and are constituted by how we go on and the interpersonal mechanisms that incorporate these practices into us and us into those practices.

Examples, as Wittgenstein says in *On Certainty*, are an important part of establishing a practice, of instilling in learners what it is to engage in that practice aright. The examples at some point will work to bring it about that a learner knows how to go on; their force is not exhausted or expressible by a rule, which would be subject to all the skeptical problems about underdetermination already rehearsed here. The examples help to instill the right behavior in beings enough like the teachers to share their form of life, to get the point of the practice, to come to grasp how to go on: they are the voice of the practice.

If this account is on the right track, then matters seem to be looking up for clinical judgment. Rules of the sort that make up practice guidelines and the rest of the formalized model cannot be the end of the story. Those rules depend on an underlying, socially mediated ability to know how the rules, or action-guiding information of any sort, hook up to the world. We might as well call this ability judgment—in medical contexts, clinical judgment—and grant that it deserves attention on its own.

But, as with the cautions about philosophy and science, the idea that we have decisive Wittgensteinian reasons that vindicate clinical judgment as a mode of decision making worth taking seriously may still be too hasty. We have to consider the possibility that the "rest of the story" about rule following may be only tangentially relevant to the clinical contest.

Note that Wittgenstein introduces the notion of practices in order to respond to certain kinds of philosophical questions that threaten to seduce philosophers into metaphysical excesses. He is absorbed by questions such as, "How is it that a set of marks, or a series of sounds, are not merely dead inscriptions or simple noises, but come to have a meaning at all?" Philosophy has suffered from a temptation to say that meaning is caught up with some special mental power—"*certain definite* mental processes bound up with the working of language, processes through which alone language can function" (BB, p. 3)—but that is simply obfuscatory. There is a perhaps more sophisticated temptation to say that it is the use of those marks according to rules that does the trick, but as a complete account, that induces by-now-familiar cramps: worrying about whether there is an indefinite series of rules, each specifying the meaning of a previous rule. Where does this all end up? In what we do without further interpretation, what we are trained to do (or to see, or to feel), in our mastery of certain repeatable techniques, in our practice. Consider Wittgenstein's builders, described in the first pages of the *Investigations*:

> Let us imagine a language . . . meant to serve for communication between a builder A and an assistant B. A is building with building-stones: there are blocks, pillars, slabs and beams. B has to pass the stones, and that in the order in which A needs them. For this purpose they use a language consisting of the words "block," "pillar," "slab," "beam." A calls them out;—B brings the stone which he has learned to bring at such and such a call.—Conceive this as a complete primitive language. (PI § 2)

Like any other, this language could be described in terms of rules. "Slab" means "bring an object of *this* sort from *here* to *there*." But despite the primitiveness of the builders' language, this rule is just as subject to the indefinite permutations of interpretation and reinterpretation as any other. The next time the builder says "Slab," the assistant might bring a beam, or take away a slab, or smash slabs into rubble—and any of these actions could be regarded as governed by a consistent understanding of the "Slab" command. ("Slab" means "bring a slab except when six slabs have already been delivered during any one morning"; "slab" means "bring slabs except on Leap Year Day; then, smash slabs into small bits"; and so on.) Among the builders, then, there is just as much need for example and training in the establishment of a practice as in more complicated language games. There are just as many loopholes to fill.[11]

But these reflections do not seem to provide any support for judgment models over formal decision-making models. Both models must rely upon some kind of training and example imparted at some point in the learner's life in order for him or her to be able to get off the ground. And that is what Wittgenstein seems to be worrying about: not which practices are better or worse than others for some end, but what is involved in our being able to engage in practices at all, and what the implications of that fact of our natural history are for language and the confusions we get into when we start thinking hard about it.

Still, what is involved in knowing how to go on as a builder's assistant and knowing how to go on as a physician seem to present us with strikingly different orders of complexity. The multiple ambiguities that the builders face are a function of a particular way of looking at the relationship between language and the world, one that focuses on the propositional content of rules in order to provide them with sufficient fixity. But the ambiguities faced by people in medical contexts do not emerge solely from a bad philosophy of language. And about these less philosophical forms of ambiguity Wittgenstein thought mistakes were also likely, if we did not attend appropriately to the importance of a kind of training that is not exhausted by giving people explicit rules.

Correcter Prognoses

Consider Wittgenstein's explicit discussion of "expert judgment" in Part 2 of the *Investigations*:

> Is there such a thing as "expert judgment" about the genuineness of expressions of feeling?—Even here, there are those whose judgment is "better" and those whose judgment is "worse."
>
> Correcter prognoses will generally issue from the judgments of those with a better knowledge of mankind.
>
> Can one learn this knowledge? Yes; some can. Not, however, by taking a course in it, but through *"experience."*—Can someone else be a **man's** teacher in this? Certainly. From time to time **he** gives **him** the right *tip.*—This is what "learning" and "teaching" are like here.— What one acquires here is not a technique; one learns correct judgments. There are also rules, but they do not form a system, and only experienced people can apply them right. Unlike calculating-rules. (PI IIxi, p. 227^e)[12]

This sounds like just the sort of thing that judgment proponents have in mind, and just the sort of instruction they espouse. One must note right off, of course, that Wittgenstein's "prognoses" did not concern the pathologies operating within people, or what was likely to happen to them, or what kind of therapeutic interventions might be best. His topic is discerning the genuineness of expressions of feeling. And here may be the rub. About whether such expressions are genuine or feigned, there is nothing but human subjectivities measuring each other. About medicine, there are pathogens and drugs, tumors and radiation. And about their interactions, certain things are true and certain things are not. What is or is not the case about these interactions will have certain effects on human health and well-being; these are all matters of fact, formalists may aver, and scientific inquiry is our best way of determining just how they fall out.

But before considering these disanalogies between judgments concerning feeling and judgments concerning medicine, I want to draw attention to the fact that Wittgenstein in this passage is not directing his attention merely to philosophical doubts about how signs get meaning at all, or to matters of similar generality. Section 12 describes particular practices—calculating, judging colors, determining whether feelings are genuine or not—and in the quoted remarks Wittgenstein is clearly dealing with how expert judgments get made in an area in which no technique or system is to hand.

Consider both the practice of the expert judge of feeling and the practice of the builders and their assistants. How is acquiring expert judgment about the sincerity of feeling expression like the kind of training that the builder's assistant had to have in order to move the slabs, beams, pillars, and blocks around?

What the builders do can be expressed in terms of a system of definite rules, but the ability to use those rules presupposes that a very complex form of "knowing how to go on" is imparted to assistants. This cannot be done by someone giving apprentice builders' assistants a list of rules, but by teachers giving learners the right kind of indication, a tip, a wink. But despite the fact that the excluded interpretive possibilities are indefinite, we regard the builder's assistant as having acquired a very basic form of expertise here, a sort of "knowing how to go on" quite fundamental to participating in this complicated form of life we humans share.

In contrast, those who learn from the expert judge of sensibility—some of them, anyway—will develop something rarer and more difficult, the kind of *Menschenkenntnis* required for good judgment about people's feelings. What Wittgenstein is drawing our attention to here is the existence of human practices in which excellence is not exhausted by fully explicit rules or by the ability to follow such rules.

Our concerns about how the builders avoid getting lost in indefinite underdetermination of their rules are, so to speak, philosophical concerns, and the allusion to human practices is designed to allay them. But the judges of feeling face indefiniteness and underdetermination as a practical, not a philosophical problem. We might mark the distinction this way: the problems judges of feeling face in making their determinations, unlike those faced by builder's assistants, occur within a practice and cannot be resolved by reference to the necessity of their participation in a practice.

We might speak of the kind of judgment that is instilled by being a part of a practice—the sort builders' assistants have—as *ordinary judgment* and the kind that is refined within a practice as *expert judgment*. There are, for example, some, recipes that can be followed by the rawest, most uninspired and autodidactical of cooks; the rules so plain and so explicit that any one who knows how to read and possesses even a pinch of mother-wit will know how to go on. Mathematical operations can be like this; so is what the builder's assistants do. But there are some activities—such as judging the genuineness of expressions of feeling—that require expert judgment, judgment that goes beyond the capacities required in any practice whatsoever. In such activities, any rules that are useful will be subject to multiple ambiguities of application that are not just philosophical in character, but quite practical, and that require experience of the right sort to apply.

But what reason have we to think that medical judgments are going to be more like judgments concerning the expression of feeling than judgments concerning slabs, beams, and blocks? That is, why think that the practices required for good medical judgments are those presupposed in the morning-report model—expert teachers inculcating neophytes into a practice by giving them lots of the right tips—and not those in which the background of practices, of trainings, of shared ways of going on required for the use of the appropriate rules is, as it were, much further back in the learning history of new physicians? After all, while we lack a system of rules for judging sincerity, we seem well on the way to developing one for treating cardiac disease.

Part of the answer would seem to lie in the consideration that just as the ability to be guided by rules at all will require a sort of generic training in order for creatures with our natural histories to respond correctly, so rules of specific sorts will open themselves up to specific forms of ambiguity and indeterminacy, requiring training of a special sort to turn otherwise capacitated language learners into people who know how to go on in the right way in the specific domain to which those rules are pertinent.

What will it take to keep the ambiguities in applying clinical guidelines to clinical situations under control, so to speak? Clinical judgment has clearly been more like judging feeling than doing straightforward arithmetic or making a cake from a mix—a matter of expert, not ordinary, judgment. However, what purports to be such a system for medicine is on the way. Is there anything in Wittgenstein's ideas that supports spending time, effort, and dollars exploring the judgment model explicitly—about, for example, seeking out and trying better to understand the "superstars of judgment"?

There is, of course, this practical concern: if we assume that the formal model will continue to flourish and that the evidence on which it stands will grow in reliability and credibility, then whole regions of medical practice may become areas in which only ordinary judgment is needed to operate well. But given the burdens of doing good outcomes research, and of negotiating the epistemic and political currents between such research and the production of good practice guidelines, it is extremely likely that cutting-edge areas of medicine will remain for which guidelines do not yet exist or are only partially based on convincing evidence. Further, it is surely always possible that a particular case, even in areas for which good guidelines exist, will toss up important features in ways for which formal evidence-based techniques provide no precedent, just as people can express feeling in unaccustomed ways; the person of good judgment in either area will be able to make sense of the novel.

But there is also an epistemically more interesting possibility: practicing with guidelines might itself make visible new distinctions among patients and among diagnostic and therapeutic strategies that become patent to those familiar with the guidelines—but who also have the ability to look through and beyond them for new clinical saliencies to which the guidelines are not sensitive. That is to say, the formal evidence model might be used to enhance, rather than to replace, clinical judgment. The intuitive idea is this: if clinical guidelines can hold a number of the relationships between symptoms and therapies constant, so to speak, clinicians will have a less cluttered perceptual field with which to deal; new, more subtle kinds of clinically relevant indications may stand out more sharply if there are fewer unruly phenomena to which to attend. In possession of a rule that allows you to identify a smile, you may be in a better place to distinguish a grin from a smirk.

Suppose outcomes research were to show that piercing children's eardrums with small tubes was largely useless as a response to otitis media, for example, or that hormone replacement therapy provides women without a family history of heart disease only a marginal extension of life expectancy. Such results might put clinicians in a position to attend with more imagination and sensitivity to other considerations that bear on the well-being of a particular child with an earache, or on how a particular woman is experiencing menopause.

But this sort of scenario, of course, presupposes that physicians remain more like judges of feeling than like builder's assistants. And this presupposition might strike formalists as dangerous: if part of the purpose of the formal approach is to cut down on differences in practice, to standardize what physicians do, then allowing them to use guidelines as a perspective from which to be able to perceive subtle but clinically relevant distinctions between the patients grouped together by the guidelines seems to defeat the purpose. At the same time, the idea that guidelines themselves might serve as a jumping-off place for finer forms of perception and better-informed judgments seems a hypothesis worth exploring, given the different ways diseases can play out in different people's lives.

Both the practical point that we can expect quite a wait—to say no more—before medicine is fully standardized and the epistemic point that guideline-informed practice may provide a basis for further insights into how to respond to particular situation suggest that clinical judgment will become an area of more explicit and intense investigation. Scriven's suggestion about studying the superstars of clinical judgment gets an interesting variation. How might they practice with guidelines at their discretionary disposal? And what would be the outcome of their doing so? Better medicine? More cost-effective medicine? One of these, but not both? Neither?

Wittgenstein does not answer these questions. But he does provide some pertinent thoughts—warnings about how methodological enthusiasms can blunt careful perception of particulars, and reminders about what expert judgment is and how it can work—that can make the questions important to us. Expert judges, not enraptured by scientific method but with formal rules to hand, might do still better than either expert judges on their own or those of only ordinary judgment whose reliance on the rules is more determinate. We would do well to explore such ideas before we transform medical education and practice into an area where ordinary judgment is all we aspire to.

Notes

Citations to Wittgenstein's texts will be given in context, using the usual abbreviations for his works. Carl Elliott end Hilde Lindemann Nelson were very generous with the time and trouble they spent trying to help me figure out "where to go" with this essay.

1 It is worth mentioning that utilitarianism's aim to make moral decision making calculative is solely a matter of aspiration. In many of the kinds of difficulties in which we might want some help from a moral theory, determining which course of action would bring about the best overall consequences will require extremely good judgment.

2 The discussion of this point is found in Michael Scriven, "Clinical Judgment," in *Clinical Judgment: A Critical Appraisal*, ed. H. Tristram Engelhardt, Jr., Stuart F. Spicker, and Bernard Towers (Dordrecht, The Netherlands: Reidel, 1979), pp. 5–6.

3 Ibid., p. 3.

4 I hasten to add that this is a wholly suppositious example.

5 For a discussion, see J. E. Wennberg, "Dealing with Medical Practice Variations: A Proposal for Action," *Health Affairs* 3, no. 2 (1984): 6–32.

6 See the essays in Philip Boyle, ed., *Getting Doctors to Listen: Ethics and Outcome Data in Context* (Washington, D.C.: Georgetown University Press, 1998).

7 Of course, even were this so, it might not be a decisive moral objection; the cost effectiveness of "formalized" medicine might allow a more just distribution of medicine of decent quality, even if such policies lowered the chances that economic elites would receive excellent-quality medicine. But my concern in this essay will be primarily epistemological, not moral. Proponents of the formalized approach claim that their approach will be epistemically superior as well.

8 For an interesting (and risible) statement of this view, see Marx Wartofsky, "Clinical Judgment, Expert Programs, and Cognitive Style: A Counter-Essay in the Logic of Diagnosis," *Journal of Medicine and Philosophy* 11, no. 1 (Feb. 1986): 81–92.

9 Scriven makes just this point in "Clinical Judgment."

10 According to Charles Larmore, *Patterns of Moral Complexity* (Cambridge: Cambridge University Press, 1987), to whom I owe this point, the discussion in question occurs "just before the chapter on schematism" (p. 2).

11 For a clear and sophisticated discussion, see Barry Stroud, "Mind, Meaning, and Practice," in *The Cambridge Companion to Wittgenstein*, ed. Hans Sluga and David G. Stern (Cambridge: Cambridge University Press, 1996).

12 I follow the convention introduced in Judith Genova, *Wittgenstein: A Way of Seeing* (New York: Routledge, 1995), of graphically marking Wittgenstein's use of masculine marked terms as generic denoters, to, as she says, "underscore the point that not all native speakers or philosophers are men" (p. xiii).

5   Wittgenstein's Startling Claim:
Consciousness and the Persistent Vegetative State
Grant Gillett

"Consciousness is as clear in his face and behaviour
as in myself."—Wittgenstein[1]

Many people worry that patients in a persistent vegetative state (PVS) may have an active conscious life to which we have no access. After all, PVS patients experience sleep-wake cycles, may open and close their eyes, even appear to smile or grimace. Family members and health care workers worry especially that such patients experience pain.

Yet Wittgenstein claims that we can recognize consciousness in another as clearly as in ourselves. This is a startling claim to anyone who has been bewitched by the Cartesian tradition of the skepticism of other minds; I will therefore refer to it as "the startling claim." If Wittgenstein is right, then we can see whether a human being is conscious, and our worries that PVS patients may have a conscious life are unfounded. Even if this is only typically and not invariably the case, it is an important claim because it allows us to decide whether there are exceptions to the general rule and what conditions should lead us to suspect we might have an exception before us.

In this essay I begin with an outline of the Cartesian view of consciousness and mental life that sets up the problem for understanding PVS. Next comes an argument aimed at undermining the Cartesian view of consciousness and the skepticism of other minds that springs from it. I then present a positive account of Wittgenstein's view of consciousness; a clinical characterization of consciousness congenial to Wittgenstein's view; a view linking consciousness and concept use; and an argument linking concept use and cortical function. Finally, I draw from these Wittgensteinian thoughts some implications and conclusions for PVS.

The Cartesian View of Consciousness

The startling claim is startling because most of us are closet Cartesians. That is, we believe the subject has privileged and direct access to consciousness, that others have only indirect access, and indeed, that consciousness can be intact even though inaccessible to others. Because the Cartesian theory is central in the abiding worry about PVS, it is with that theory that I shall begin.

Cartesian conceptions of the mind accept the premise that my thoughts are in my head and essentially private to me, and that yours are similarly internal and private to you. I can, however, communicate my thoughts to you using language. Therefore, the theory holds, with Frege, that my "inner thought" gives meaning to my words, and your inner thought is the basis of your understanding of my words.[2] But how is it that I know what you mean? I cannot see what thoughts accompany your words. I cannot assume that my understanding bears anything more than a fortuitous relationship to your meaning, because there is no essential relationship between my thoughts and yours. On the Cartesian view, we have no way of detecting the meanings that lie behind our words, and therefore no way of knowing what to make of what another person says.

This, however, is a disturbing conclusion, because it implies that we have no justified expectation of achieving a shared conception of truth. To be justified in such in expectation, we would need to have some reason to believe that you and I mean the same thing when we say something. Consider, for example, a remark such as, "That cat is black." On the Cartesian view, this remark is given its actual meaning by the thoughts it expresses. Thus, for all you know, I might have the thoughts that go with your words "the dog is white" when I say, "The cat is black." This implies that what I say or think could be true, while what you say (or think) using the same words is false. Frege argues that this is an intolerable conclusion, because scientific knowledge is built on the presumption that what is true is true for anyone who thinks it, and that the truth can be communicated and understood by different thinkers. If, however, what I think when I report a scientific result or write an article like this is quite possibly opaque to you, and vice versa, because neither of us can ascertain the connection between words and their accompanying inner states, then a universal foundation for scientific inquiry is undermined.

For this reason, Frege claims that the communication of thoughts, and therefore the determination of truth, cannot depend on what he calls "men's varying states of consciousness." [3] If truth depends on inner states, then our apparent store of objective knowledge about the world can be based on nothing more than guesswork about the thoughts behind the words of others, and there may be no essential connection between what I hear and what you mean.

This view is, of course, not meant to be taken seriously; it is as absurd as it is unavoidable on the Cartesian model that informs our common conceptions of consciousness and its mysteries. Therefore, we need an alternative understanding of consciousness, one that explains the fact that we can communicate, without systematic ambiguity, about contentful conscious experience. Such a need makes it worth the trouble to visit (albeit briefly) the intellectual bloodbath surrounding the private language argument.

The Private Language Argument

In the section of the *Philosophical Investigations* referred to as the private language argument, Wittgenstein draws together strands concerning the argument that meaning is use, the idea of rule following, and the central importance of "agreement in judgements" (PI § 241) in any theory of meaning and truth. [4] What Wittgenstein seeks to establish is that it is not plausible to believe that a person can speak, reason, and have thoughts about objects that are experienced only in their own private conscious world. And if mental contents are not private, then there is an essential and typical link between consciousness and the public sphere.

MEANING IS USE

Wittgenstein's first major claim in the *Philosophical Investigations* is that understanding what a word means is a matter of correct use. He remarks, "For a large class of cases—though not for all—in which we employ the word 'meaning' it can be defined thus: the meaning of a word is its use in the language" (PI § 43).

Thus, if someone were to say "Get me the book!" and I got the newspaper, there would be some doubt about my understanding of the term "book." If the person asking me wanted to check up on my understanding of the term "book," then she might ask, "Do you know what a book is?" or ask me to tell her whether or not certain things she showed me were books. If I failed to demonstrate any familiarity with the use of "book" in real-life situations such as these, then she would be entitled to conclude, from my failure to use the term properly, that I did not understand it. Thus we could say, with Wittgenstein, that the criterion of understanding is correct use, and that in the general run of things, that is a public fact.

Contrast with this view the Cartesian picture of inner ideas or images. To see how the Cartesian view fails, consider a person whose understanding of "book" is in question and who says, "I am trying to understand what you mean by 'book' but I can't concentrate because this image of a thing with two covers and pages inside it keeps distracting me." Obviously, the person has an appropriate image in their head, and, ex hypothesi, the image comes to mind when the term "book" is mentioned. But he does not know how to use whatever is occurring to him, and therefore he cannot be said to understand the meaning of "book." By contrast, he would be said to understand if, in the relevant circumstances, he responded in the right kind of way or exhibited an ability to make correct use of the term "book."

RULES

When we reflect on what is required to latch on to the meaning of a word or phrase, it becomes clear that there is a set of informal and perhaps inchoate rules that must be obeyed. Among other functions, the rules determine what kinds of things the word should be applied to and what linguistic links fill in its meaning. Obeying a rule, I would argue (with Wittgenstein), is a matter of mastering a practice or technique (PI §202). Such mastery is gained by using the relevant term in such a way that one's mistakes and successes are manifest and can be corrected, usually through communicative interaction with others.[5] Wittgenstein concludes, "Hence it is not possible to obey a rule 'privately': otherwise thinking one was obeying a rule would be the same thing as obeying it" (PI §202).

For us, the important claim is that a person must have some sufficiency of correct application and discursive knowledge to count as having understood a concept. Thus, the cognitive or discursive grasp of any rule is predicated on the actual use one is able to make of it in practice. This is important, because it suggests that a key aspect of thinking—understanding the meanings of words—is something that is tied to the public domain. It also suggests that the meaning of any word, such as "conscious," is determined by something public. The public nature of meaning brings me to my third point in setting up the private language argument.

AGREEMENT IN JUDGMENTS

When does a person count as having mastered the rule-governed ability to make use of a term? When he or she has achieved a sufficient level of agreement in judgments with others who have mastered the relevant ability or practice. For instance, I would count as a person who grasped the concept *square* when I was able reliably to judge *that figure is square, this is not square, a square is a four-sided plane figure*, and so on. Such an ability, evident in my use of propositions containing the term "square" would both constitute, in part, and exhibit a grasp of the concept *square*. Thus my understanding, or grasp of a concept, goes hand in hand with exchanges in which my judgments are manifest to others and corrected by them so as to correct me when I am wrong. The result of this process is that my judgments gradually converge with those of others who already understand the concept I am trying to learn.

These reflections tying understanding and meaning to rules, and thereby to communicability and corrigibility, have important implications both for our view of what it is to have contentful conscious experience and for our use of mental terms such as "anger," "seeing something red," "being in pain," and "conscious." First, they imply that conscious experience and the categorization or conceptualization of its contents are tied to my participation in a public milieu where I have learned rules governing the use of concepts. As a corollary of this, my use of concepts to make sense of the experiences I have is built upon my being with other minds and their interactions with me. We shall see that this, in turn, leads to a fairly strong endorsement of Wittgenstein's startling claim and therefore that it touches directly on the abiding worry about PVS.

Wittgenstein on Consciousness

Wittgenstein had some very instructive things to say about consciousness and the mind. They make his startling claim plausible and give us a basis on which we can say something about what it is possible to know about patients in PVS and their mental states (or lack thereof). One or two preliminaries are in order.

First, one can be *conscious simpliciter* or *conscious of* this or that thing. An act of being conscious of something necessarily takes an object; that may be an external object, a state of affairs in the world, or one's own state of mind. It is therefore intentional in Brentano's sense. I will argue that *being conscious of something* or *consciousness as intentionality* is the basic type of state, and that consciousness simpliciter is merely a conglomeration of acts of being conscious of this or that. (In this I agree, reservedly, with Hume when he claims that "with regard to the mind, that we have no notion of it, distinct from the particular perceptions." [6])

Corrigibility and communicability imply that our language about mental attributes such as consciousness shares a certain feature with all meaningful words. That feature is that the meaning of any word depends on more than one person knowing whether it is being used correctly on a sufficient number of occasions, so that the requisite "agreement in judgments" to establish rule-governed use is achieved (PI § 241.) This in turn implies that one cannot learn the meaningful use of a term like "conscious" unless there is some way that correct use of that term can be established intersubjectively. Thus, my saying of you that you are conscious (and therefore, if you are learning the concept, your use of "conscious" to describe yourself) must be based on something that can be agreed on by both of us and not merely on my guess about hidden goings-on in your head.[7] This in turn implies, among other things, that consciousness is a property of beings about which we can make judgments and not a feature of internal processes and states about which we can only conjecture. Wittgenstein emphasizes this point when he remarks, "Look into someone else's face, and see the consciousness in it, and a particular shade of consciousness. You see on it, in it, joy, indifference, interest, excitement, torpor and so on" (z § 220).

Of course, we may not always be sure whether a given individual is conscious, but we must, in principle, be able to agree on what would count as being in that state for the word "conscious" to have a determinable meaning. Thus, an adequate account of consciousness should be able to say how, in general, we are quite competent in judging whether self or others are conscious. At this point there is an interesting convergence between Wittgenstein and clinical neurology.

Clinical Neurology and Consciousness

One crucial indicator of danger in a patient with a head injury or other intracranial catastrophe is deterioration in that patient's level of consciousness. The necessary observations must be made by several different third-person attendants and form a coherent record such that the observations of different observers are commensurable. Thus, it is central to neurosurgical practice to be able to monitor and record the patient's level of consciousness in some reliable and repeatable way. If Wittgenstein is right, and we can observe the consciousness of another person, then this would be possible. In fact, he prefigures the idea that it might be possible when he says that we give *signs* of our mental states (z § 515). A doctor or nurse relies on such signs.

When a doctor or nurse assesses the consciousness of a patient, he or she normally uses the Glasgow Coma Scale, which gives scores for different levels of consciousness.[8] She assesses whether the patient's eyes open spontaneously (4), to voice (3), to pain (2), or not at all (1). She also rates his motor response, which may be appropriate and to command (6), a directed response to pain (5), a semi-purposeful response to pain (4), a flexor response (3), an extensor response (2), or no response at all (1). A further assessment is made of his verbal ability, which may be rational/sensible (5), irrational/confused but articulate (4), incoherent words (3), moaning and groaning (2), or no ability at all (1). The scale thus runs from fifteen points to three points, and it depends on the fact that consciousness is manifest. What is more, the upper reaches allow us to discriminate degrees of consciousness.

Imagine two patients: one, A, is awake and alert, answering questions, doing what is asked of him, and taking note of his surroundings (GCS score 15); the other, B, is intermittently opening his eyes to command, moving only to ward off painful stimuli, and producing occasional bouts of swearing (GCS score 11). The first patient is unequivocally conscious and the second is not fully conscious. (Here I am just reporting a clear neurological distinction measured by the difference in scale scores). What is this differ-

ence between the two predicated on? It seems to me that the crucial difference lies in the cognitive engagement between the subject and his environment. It is clear that A can direct his cognitive abilities on any aspect
of his environment, whereas B cannot. Of A, one could say that his
thought or understanding can, depending on his level of conceptual development, engage with anything within his perceptual range. The responsiveness evident to observers shows a level of complexity in the way
he can extract information from what is around him and, by so doing,
appreciate its qualities—color, warmth, softness, movement, sounds,
friendliness, and so on. Therefore, he is in a state that enables him to
think about what is happening to him and incorporate these thoughts
into a growing appreciation of himself and his present situation. (We
could call this a state of potentially differentiated intentionality, or object-
directedness, thereby connecting these present thoughts to Brentano's
characterization of the mental domain.)⁹ The cognitive state of the patient entails that he can "tell" or report on what it is he is perceiving, because his verbal abilities are just a subset of the range of cognitive abilities
he exercises with respect to what is around him.

We might conclude, when we look carefully at the neurological instru-
ment we use to document consciousness, that we use signs of cognitive
engagement with the world and attend closely to the manifestations of
cognitive attunement to the environment. Here it is worth taking note of
Wittgenstein's remark, "An inner process stands in need of outer criteria"
(PI § 580), because it seems that when we refer to a person as conscious, it
is the outer criteria that tell us whether the inner state is present in exactly
the way he suggests. From this we can conclude that when I say, "I am
conscious," I am, in effect, saying, "I am in the state that is generally referred to as consciousness." Wittgenstein's view is thus congenial to the
view we might adopt on the basis of clinical facts: that we can achieve
inter-observer reliability in assessing levels of consciousness in head
injury.

It is because of the externally directed and animated nature of a person's
engagement with the world that we can say with confidence that he is
conscious in that he fulfills the criteria that underpin our communicable
and corrigible use of the term "conscious." When a person says he is conscious he is saying that he is in the state that justifies anybody, including

himself, in using the term "conscious." (He could say this even though he may have ways of knowing that he is in the state he is in that are different from the outer criteria used by the rest of us [z §472].) It also follows from this view that our assessment of someone's level of consciousness relates closely to the attentive and responsive abilities toward things around us that, I have argued, are central in the grasp of concepts.

Of course, it is not invariably the case that an observer can tell that someone is conscious, and we might expect that failure of discernment to occur whenever some internal abnormality in the body has disconnected cognition from its normal manifestations. However, at this point I will regard it as plausible (if not proven) that consciousness is a matter that is clear to observers in the majority of cases, so that we all do use the term "conscious" in a corrigible and hence reliable way. We must now see whether an account of significant or contentful consciousness can be related to what we know about brain function, so as to determine which lesions are likely to affect consciousness and thought in the ways we believe it to be affected in pvs. If we can do that, then we can use the resulting neurophilosophical insights to assess our abiding worry about the pvs patient.

Consciousness and Concepts

I have argued that being conscious of something—say, a car—is being cognitively engaged with it in some particular way. In fact, I am always conscious of something as a certain kind of thing—a big noisy moving thing, a car, a 1956 Pontiac—and this has implications for the ways in which the conscious subject responds to the thing in question. But to distinguish something as a noisy vehicle, a car, or a 1956 Pontiac is to characterize it according to certain concepts. I have argued that the ability to conceptualize something is essentially linked to articulated rules for the use of signs (typically, though not exclusively, tied to the meaning and understanding of language). This, in turn, implies that the contents of consciousness, or consciousness as a complex of instances of being conscious of this or that, involves high-level informational articulation and integration in the central nervous system (cns). This, in turn, suggests that the contentfulness of consciousness is a function of complex conceptual abilities and that when the cerebral cortex is lost, then consciousness is also lost. However, we need to take this set of arguments at a slightly less breakneck pace if they are to be convincing.

The first constraint on any account of consciousness or mental content is the principle of economy. This requires that we not credit any subject with mental abilities that go beyond those needed to account for the behavior we see. Thus, for instance, we would not say of a snail that it is trying to hide in the hedge when its behavior is perfectly well explained by a taxis toward dark or shady places. (We might, of course, be able to devise a number of experiments that showed us exactly what the movement was aimed at.) In order to justify crediting the snail with thoughts about hiding in hedges, we would need evidence that it had a concept of hedges as distinct from other shady places such as trees, shrubs, or walls, or hedges as distinct from green leafy masses, and even then we could not possibly get it to show the necessary distinction between artifactual and natural arrangements of shrubbery, boundaries, and so on, all of which would go into thoughts about hedges. We just could not find evidence of this level of differentiation in the snail's behavior, and thus we would be wrong to think of the snail as intending to hide in the hedge. We are driven toward a cognitive account of consciousness when we try to make meaningful and defensible distinctions—in the light of the principle of economy —between conscious and mere reflexive or respondent information-processing systems. Wittgenstein remarks, "I only use the terms the expectation, thought, wish, etc., that p will be the case, for processes having the multiplicity that finds expression in p, and thus only if they are articulated. But in that case they are what I call the interpretation of signs. I only call an articulated process a thought: you could therefore say 'Only what has articulated expression.' (Salivation—no matter how precisely measured—is not what I call expecting.)" [10]

Wittgenstein's contrast between salivation and expectation emphasizes the vast difference between a conditioned reflex and the use of a concept in thought. Whereas the former is a rather narrowly defined response rigidly linked to a stereotyped or specifiable trigger, a concept is an exploratory and unifying tool that can be employed in many diverse situations according to our interests (PI § 570). It is useful precisely because it ranges over a wide variety of situations and potentially links experiences that may otherwise have no obvious connection with each other. It does this by framing those situations against a conceptual system that is finely crafted to detect and exploit any aspects of an experience that may relate to the purposes of the concept-user. A thought is articulated in terms of the concepts making it up. (For example, "that frog is green" involves the concepts *frog* and *green* and therefore is articulated in terms of the complex patterns of links upon which those concepts draw.) What Wittgenstein calls the "multiplicity that finds expression in p" (where p stands for some specification of content in terms of concepts grasped by the thinker) is the richness of the links expressed in the semantic components of p. Thus, a conscious mental act such as an "expectation," with its rich and open-ended connections linking it to nuance and conjecture, clearly differs from "salivation," with its simple and circumscribed nature.

To fill out this multiplicity or complexity is to spell out what it is to be conscious of some entity or item, such as, for instance, a stoat, a bat, or a matchbox. The complexity of consciousness is apparent in any of these cases. One can identify the stoat as a stoat only if one also can think of it, for instance, as an enduring object, as an animal, as able to perceive and act on what it perceives, and so on. To think of a matchbox is related to thinking of the things it normally contains, to thinking that it can be opened, and so on. Abilities to detect these features of an object of consciousness are established over time and in association with the person's mastery of the ways of thinking connected to each of the conceptual terms. I have already argued that the ways of thinking are themselves mastered in interactive human activity. For my present purpose, the interplay of directed gaze, investigative activity, and flexible responses to an object underlie the conceptual complexity of human thought. In fact, any articulated system of concepts involves just those abilities to correlate a wide range of experiences and analyze them according to different plans and projects that characterize conscious rather than reflex activity. Thus, consciousness and conceptually contentful experience go together, and consciousness simpliciter is a state comprising innumerable acts in which one is conscious of this or that.

To summarize, I have argued that a philosophical study of consciousness suggests two things: that consciousness is manifest between subjects, and that consciousness has an essential link with concept use in that both involve selective, articulated responsiveness to things in the environment.

Cortical Function and Consciousness

The cerebral cortex is the convoluted layer of complex neural circuitry that carries out the most articulated and detailed processing of inputs to the brain. The cortex includes the neocortex, archicortex (juxtallocortex), and paleocortex (allocortex). The paleocortex is concerned with olfaction. The archicortex includes the hemispheric parts of the limbic lobe and is important in understanding the interface between higher-level processing and more primitive mechanisms. The central structure in this system is the hippocampus, which has a role in memory formation, learning, and orientation.

In clinical neurology and cognitive psychology, we have long since determined that conscious cognitive activity can (as a gross simplification) be thought of as involving three levels of cerebral function: primitive arousal mechanisms, nonspecific attentional and orienting mechanisms, and detailed higher-level analysis of information. Cortical and subcortical function are, of course, intimately linked to the extent that many of the interwoven and dynamic patterns of cortical activity may involve subcortical relays. For the sake of simplicity, we could say that the most primitive and general reactions (sleep/wake cycles, gross turning up or down of the "gain" or salience of various inputs or systems, and stereotyped patterns of activity) are mediated in the brain stem (though not necessarily controlled there), and orienting responses to various classes of stimuli are mediated by the diencephalon in conjunction with primitive cortical systems. These more primitive structures are also vital for the relay of information to the neocortex, but when the cortex itself is destroyed, an individual enters what we call the vegetative state, wherein the only activity shown is of a reflex or vegetative type.[11] Most patients do not recover from such a state, but those who do have no recall of what went on while they were in it. Those who do not recover are said to be in a persistent vegetative state, which can only be diagnosed when the cortex is widely destroyed according to our best metabolic and functional evidence. (Vegetative states stand in marked contrast to the "locked-in syndrome" or *coma vigilante*, where the individual is cut off from motor expression by a small but critical subcortical lesion and yet, if he recovers, shows an intact [articulated and conceptual] awareness of what went on while he could not communicate.[12] I will discuss this further below.)

"Lower-brain" or brain-stem centers interface with the cortex through the limbic system, in particular the hippocampus. The limbic system functions with extensive inputs from all areas of the neocortex and is also closely connected to motivational and arousal centers (for example, in the hypothalamus and midbrain). It appears that this area enables drive-related or undifferentiated cues with novelty or potential behavioral relevance (because, say, of their role in learning history) to capture processing space by influencing the "gain" in the neocortical arrays. One could gloss this phenomenologically as being correlated with a vague premonition or "feeling" that something deserves attention, without being able to say precisely what it is. To function effectively, the mechanism responsible for this kind of presentation would need to be connected to the organism's spatiotemporal tracking abilities, to those centers providing input about drive states and their satisfaction, and to the information in memory. Luria, remarking on patients with lesions in these areas but intact higher (neocortical) processing, notes that the "potential integrity of the higher, voluntary forms of attention in the presence of a primary disturbance of its elementary forms is an important sign distinguishing these patients from those with lesions in other sites."[13] By this Luria means that although patients with lesions in these areas do not show the general bodily responses and reactions evoked by drive-related stimuli such as pain or fear, they can compensate by issuing themselves verbal instructions or some other mechanism drawing on the intact higher systems.

It should now be clear why cortical function is, for the human being, an essential part of conscious appreciation and activity. Whereas the more primitive centers may potentiate conscious activity, it is the structure of the cortex that allows the correlative and selective information processing that enables human beings and higher animals to be conscious of things. This higher-brain activity structures and informs the organism's reactions to (and activity within) the world in the way characteristic of contentful conscious thought. Here empirical and philosophical studies of consciousness converge, giving a firm conceptual footing to our settled neurological opinion that conscious activity requires both brain stem integrity and cortical processing.

I have argued that attentional selectivity and information linkage between one array and a wide range of others are the basis of conscious appreciation. The first requires the preferential enhancement of centripetal information, which is a function of the reticular formation but is influenced by input from the neocortex. The second involves the complex multitracked and cross-linked processing patterns evident in all cortical areas, but particularly in the higher association areas. Stimulus-response rigidity (as in reflex activity) or automatism and lack of responsiveness (as in the partial complex seizures of some temporal lobe epileptic patients) both show, in different ways, that the behavior concerned is not conscious. Reflexes are stereotyped, stimulus-linked movements; some are evident even in a brain-dead human being. And although reflexes may enter into conscious behavior, they are not distinctively conscious in the way that makes human mental life special. Automatisms occur when a patient carries out an apparently well-coordinated and purposive set of actions but shows no appreciation of what is going on around him and, when this state (called the ictus) is over, has no awareness of anything other than having experienced a "funny turn" or a "blackout." In neither case is the activity exploratory, open-ended, and articulated with experience in general. It is this fact that makes us unwilling to endorse the idea that reflex or epileptic activity is fully conscious.

Thus, the information arriving in the brain is sorted and responded to according to a wide-ranging set of interactions between any given unit of information and the distillate of cumulative experience. New information is, as it were, linked to all the information collected and organized by the behavioral history and categorizing structures acquired by the organism. In human beings one would expect to find a pervasive effect of "the codes of language" in the organization of such structures.

The neurophysiology is relevant to the way in which we think of the monitoring that, according to Weiscrantz, characterizes conscious awareness: "it enables the animal to think about his own behaviour and others' behaviour, to communicate if it has words, as we do, so as to link objects and events in time and space that would otherwise be separated; in other words, to think and to reflect." [14]

The picture of monitoring that results is one not of a localized "second-order" processing unit that keeps tabs on the rest of the brain (and continually threatens to become a homunculus), but of a wide-ranging connectedness in which an item is articulated within the structure of information processes that underlie conceptual thought. To be able to consider information in the light of this or that other stored or current information and as affected by this or that operation is to monitor it or be conscious of it rather than merely to receive and react to it. Thus, monitoring is important because it enables flexible and goal-directed processing in the light of information distributed throughout the cognitive system. This view suggests that there is no central locus of conscious integration, but rather that consciousness is a holistic activity exemplified by a human being with whom we can interact and establish communication.

At this point it is useful to reflect on the convergence of clinical practice with Wittgenstein's views. The holistic and many-faceted reactions and responses that go to make up being conscious of a thing are exactly the kinds of patterns of behavior that we are used to recognizing in others but bad at describing. They are the kinds of patterns that ground attributions to the other based on how the other strikes us, rather than on a checklist of presented features. For instance, I might say, "He strikes me as being shy," or "It seems that has a romantic interest in her." If I were asked to justify my statement, I might say "intuition," or "I bet I'm right." In this type of case, we pick up on features of our environment that we have been learning to read since we were a few hours old. It is therefore unsurprising that they are as little susceptible to justification as statements like "That is red" or "I smell milk." These things we know, but not by a process of inference and reason. The ability to detect such things is, in a sense, inscribed in us by experience and not usually learned by conscious effort. Wittgenstein's startling claim begins to look more plausible. Consciousness may be the kind of thing we can *just see*.

Implications for pvs *and "Locked-in Syndrome"*

The implications for the pvs patient are clear but, I think, neither obvious nor unproblematic. The first consideration that ought to give us caution is that the patient in pvs has suffered a brain injury and therefore may not show the connectedness and integrity required for our normal practices of observation and mental ascription. This is an important point, but before we accept that it is a global defeater for the Wittgensteinian account, we ought to look at the type of brain damage involved and the likely effects of that type of brain damage, given our understanding of the natural history of human beings.

In PVS the higher levels of the brain have been devastated, usually by a combination of shearing stress injury to neurons and anoxia. Both of these processes selectively damage the cortical system and its ramified connectedness within itself and to lower centers. The result is that there are no longer enough megabytes, so to speak, to do the work required for conscious experience. As I have argued, consciousness depends not on a single locus of activation in the brain but on the holistic functioning of widespread and articulated loci of information processing. This includes selective attention and directedness toward the objects, conceptualized experiences, and feelings that make up conscious mental life. The richness of that activity entails that processing carried out in relatively primitive centers such as the brain stem or isolated subcortical areas does not even get close to conscious activity in the subject concerned. We can therefore say that a patient in PVS no longer has a mind: there is insufficient and insufficiently integrated activity in the brain to support those functions we call mental. The person is not *conscious simpliciter* because he or she is not in that complex state that comprises the many acts of being conscious of the things that go on around him or her.

However, the locked-in syndrome is a problem case for our normal epistemic practices. In this situation, the patient is conscious but cannot signal it by any responses or activity. Not only has he lost the ability to voluntarily direct any of his bodily activities (except, in most cases, the ability to move the eyes up and down), but also the cortical areas of his brain are disconnected from their normal bodily manifestations. This state therefore removes all the normal bodily cues that go with widespread cortical activity and makes the person inaccessible to others unless they realize his predicament and use eye movements to communicate. They are then able to establish his consciousness of what is going on around him.

Yet we might ask, "If consciousness is essentially a matter of responsiveness, how can it be that we can even think of consciousness in *coma vigilante* or locked-in syndrome?" Such patients are conscious but (largely) unable to respond. To resolve this paradox, we need to take note of the fact that Wittgenstein establishes the meaning of a term through its use in a sufficient range of typical or paradigmatic cases. Where the natural conditions underlying that practice are undermined, then he would concede that our normal criteria may not work. Thus, if I normally judge that something is a cat and therefore an animal on the basis of it being a warm, furry, purry creature that looks like a cat, then I will go seriously wrong in a world where robocats who are also warm, furry, and purry abound. In the same way, if consciousness is typically and criterially ascribed on the basis of a cns that is intact, then various situations where the cns has lost its integrity will undermine the normal criteria for consciousness. But this does not take anything away from a practice in which the state of responsiveness and so on is what we are talking about when we talk about consciousness. It just implies that there are conceivable conditions in which the state of consciousness is unrecognizable to everybody but the subject concerned.

In fact, Wittgenstein allows that mental predicates are predicates in which the basis of ascription is very different in the first-person and third-person cases (z § 472). In the locked-in syndrome, the cognitive abilities are fully and dramatically developed in the being whose embodiment is disturbed, but here, as in other cases, such as profound cerebral palsy, the inherent cognitive capacity of the affected individual is often completely misjudged because of the disturbance of bodily manifestation that we see. When we understand the aberrations concerned and take the trouble to correlate the physiology and the phenomenology, we can see exactly how it can be that the experience of consciousness is, in some sense, still present although the manifestations are grossly altered.

We come to the opposite conclusion in the case of vegetative patients precisely because we understand that the capacity to be in the type of state we call consciousness has been destroyed. In the pvs case, the cognitive or experiential core of the holistic state we call consciousness is impossible, yet some fragmentary vestiges of the manifestations remain as an expression of activity in other parts of the brain.

Conclusion

Wittgenstein allows us to see that consciousness is a state of living human beings that comprises a complex (intentional) interaction with things and events in the environment of the individual. We are good at detecting the subtle signs of this type of sensitive engagement between an individual and what is around him or her, and that is why the startling claim that "consciousness is as clear in his face and behavior as it is in myself" is true. In fact, the startling claim is commonly attested by relatives who say things like "Jim is not there anymore," or, as one woman said, "It is weird; they come in and say, 'Alan, we are going to do this,' or 'Alan, your bed needs changing,' but they are talking to nobody; he isn't there." That an ordinary person can tell when consciousness is present is endorsed by Wittgenstein's startling claim.

Wittgenstein also advises us to attend to the natural history of human beings when we make philosophical claims about the mind. Although he was skeptical about mind-brain theses in general, I do not think that we should overlook certain neurophilosophical considerations in the present case. The kind of information-processing activity that is needed for consciousness requires a high order of interconnectedness and complexity in the cerebral systems realizing it. Where these have been seriously damaged and are unlikely to be able to sustain integrated conceptual activity, then, on any broadly Aristotelian account, we can be reassured in our conclusion that the psyche of the individual person as a person has been destroyed. It seems that the abiding worry about the isolated mind in pvs can be dismissed as conceptually mistaken.

Notes

1 Ludwig Wittgenstein, *Zettel*, §221.

2 Gottlob Frege, *Logical Investigations* (Oxford: Basil Blackwell, 1977), pp. 4–5.

3 Ibid., p. 17.

4 Throughout this essay I shall use § to refer to paragraph numbers in PI and Z.

5 Grant Gillett, "Humpty Dumpty and the Night of the Triffids: Individualism and Rule Following," *Synthese* 105 (1995): 191–206.

6 David Hume, *A Treatise of Human Nature* (1739), ed. Ernest G. Mossner (London: Penguin, 1969), p. 677.

7 Grant Gillett, *Representation, Meaning, and Thought* (Oxford: Clarendon Press, 1992).

8 G. Teasdale and B. Jennett, "Assessment of Coma and Impaired Consciousness," *Lancet* (1974): 2:8.

9 Franz Brentano, *Sensory and Noetic Consciousness* (London: Routledge and Kegan Paul, 1929), p. 58.

10 Wittgenstein, *Philosophical Remarks*, ed. Rush Rhees, trans. Raymond Hargreaves and Roger White (Oxford: Basil Blackwell, 1975), §32.

11 The Multi-Society Task Force on PVS, "Medical Aspects of the Persistent Vegetative State," *New England Journal of Medicine* (1994): 1499, 1503.

12 J. R. Patterson and M. Grabois, "Locked In Syndrome: A Review of 139 Cases," *Stroke* 17, no. 4 (1986): 758–65.

13 A. R. Luria, *The Working Brain: An Introduction to Neuropsychology*, trans. Basil Haigh (London: Penguin, 1973), p. 273.

14 L. Weiscranz, "Neuropsychology and the Nature of Consciousness," in *Mindwaves: Thoughts on Intelligence, Identity, and Consciousness*, ed. Colin Blakemore and Susan Greenfield (Oxford: Basil Blackwell, 1987), p. 319.

6 Attitudes, Souls, and Persons: Children with Severe Neurological Impairment
Carl Elliott

What I want to do here is not so much to construct an argument as to confess some misgivings. My misgivings concern children of a special kind who are familiar to anyone who has spent much time around a pediatric hospital. These are children who are profoundly, irreversibly neurologically damaged. I do not have in mind children who are simply developmentally delayed. I mean children who will never be able to speak, to walk, to sit up, or to feed themselves. Sometimes they are blind or deaf. Their intellectual abilities are extremely limited, often so much so that they have never been able to recognize their own parents. The cause of their condition is often anoxic brain injury or head trauma, or perhaps, less commonly, a genetic condition with neurological effects.

What do we make of these children? How are we supposed to treat them? There is often no reason to doubt that with the proper kind of care they could live for many years into adulthood. They require an extraordinary degree of attention, and not just by health care workers. Their parents often look like war veterans, exhausted and shell shocked. Inevitably, questions arise as to how aggressively to treat these childrens' medical problems—whether to treat a pneumonia, or replace an intracranial shunt, or start dialysis, mechanical ventilation, or tube feedings. These are burdensome interventions, most of them, but interventions that can often prolong a life.

Parents and pediatricians invariably want to do what is best for the child. But when they ask me what I think would be best, I am at a loss as to how to respond. Best interests? What are the interests of such a child? Sometimes they seem to take pleasure in being stroked or being in the water; on the other hand, they often feel pain—from spastic limbs, perhaps, or more often, from the medical procedures they have to undergo. Are they loved? Yes, often very deeply. The lives of entire families are often structured around the care of such children and are marked by a special kind of grace and tragedy. Often the parents and siblings of such children have made personal sacrifices of heroic proportions, but are still haunted by guilt for what they have not done, or for the things they have secretly wished for. The irony is that the object of this guilt and sacrifice and love is totally unaware of it. How are we supposed to think about a life like this from a moral point of view? Does it make any sense to think about the "best interests" of such a child?

In his classic article "Toward an Ethic of Ambiguity," John Arras points out that when a child is devoid of any of the capacities we think of as distinctly human, asking about his or her "best interests" amounts to little more than asking which way the balance of pleasure and pain tilts.[1] Some severely impaired children will be in unremitting pain, and then we can say with some confidence that the balance tilts toward withholding or withdrawing treatment. But what about children who are not in pain, but who lack (and will always lack) the capacity to think, to communicate, to give love, or to be conscious of receiving it? The answer to the question of whether such children have an interest in continued life will often be a tentative "yes." But it will be a "yes" tinged with hesitation and uncertainty, because the question seems to skirt the more fundamental problem, which is whether the language of "interests" really captures what is morally at stake. Can we do no better than to think of these children as repositories of pleasure and pain?

Personhood and Thick Ethical Concepts

One well-traveled avenue that philosophers addressing these issues have taken is the one that leads to the question, "What is a person?" The idea here, of course, is that in order to determine the moral status of severely damaged or limited human beings, we must ask ourselves whether or not they are persons. Why? Because we all know, more or less, what the moral status of a person is—that persons deserve a certain kind of respect, that they have rights, that we owe them duties that we do not owe animals or nonsentient life, and so on. If a neurologically damaged child (or for that matter, an anencephalic, a fetus, or an adult in a persistent vegetative state) is a person, then we must treat her morally as we treat other persons. But if she is not, then we are justified in treating her in other ways—say, as a being whose interests can be safely overridden for the interests of those who *are* persons.

So how do we tell what a person is? By their capacities: intelligence, speech, self-consciousness, abstract thought, the ability to relate to others, and so on.[2] So, the argument goes, the reason we might be inclined to say that these severely damaged children are not persons is that their capacities, and their potential to develop these capacities, are so limited. If you cannot and will never think, speak, understand language, or form a relationship with others, you are not a person and therefore do not have the moral status of a person. Englehardt, for example, arguing that adults are persons because of the fact that they are free, rational, and responsible for their actions, concludes that even normal children are not persons: "If being a person is to be a responsible agent, a bearer of rights and duties, children are not persons in a strict sense."[3]

So what is wrong with this approach? Part of the problem is its (un-stated) view of the philosopher's role: as a kind of language czar, who de-vises standards for the use of words and tells the linguistic community how to use them. "Person" is thus transformed from its ordinary use to a technical, philosopher-defined use, which, on Englehardt's conception, does not include children. But the point of this kind of definitional exer-cise, while widely accepted (at least implicitly) in analytic philosophy, is by no means clear, nor is the exercise practically useful in any obvious way. In any case, it is subject to the kinds of criticisms that Wittgenstein levelled against philosophical theory: "It is not our aim to refine or com-plete the system of rules for the use of our words in unheard-of ways" (PI § 133).

A more damaging problem with this approach, however, is the way it confuses what Wittgenstein would call the "grammar" of the word "per-son." It suggests that "person" is a factual term, one that can be answered by looking at a being's capacities or potential for developing those capaci-ties. But then it assumes that the answer will give us moral guidance: tell me if this is a person, and I will tell you what you ought to do. This expec-tation is, however, deeply confused. "Person" is a moral term, not just a factual one, and the question "Is this being a person?" is itself a moral question.

"Person" is what philosophers such as Bernard Williams would call a "thick" ethical concept, one that represents a fusion of fact and value.[4] Thick ethical concepts are unlike the very general ethical concepts such as *good* and *ought* upon which moral theorists have tended to concentrate, and whose meanings are almost entirely evaluative. But they are also unlike purely factual words such as "curly," "purple," or "carrot." They have both factual and evaluative elements; that is, they fuse "is" and "ought." Thick ethical concepts like *coward, bully, cruel*—and, so I would argue, *person*—map onto things in the world, like purely factual concepts, but they also represent ethical evaluations of those things onto which they are mapped. In other words, to call a person a coward or cruel is to pick out and describe something about him, in the same way as calling him a father or a Canadian, but it is also to communicate a particular type of ethical evaluation. And the ethical evaluation is not something that follows from the description, nor is it a kind of add-on, or attachment. It is built right into the meaning of the word.[5]

This is the reason it is misguided to think we can decide what beings are persons by purely factual criteria. To try to devise purely factual criteria for applying the word "person" is like trying to devise purely factual criteria for applying the word "coward." Criteria for being a coward that made no mention of any kind of ethical evaluation—that did not convey that cowardice involves a failure of courage, that to call a person a coward is to insult him, that to behave like a coward is something that one ought not to do, and so on—such criteria would not be able fully to convey the meaning of the word. Indeed, it is hard to see how one could understand how people apply a thick ethical concept without sharing (if not actually, at least imaginatively) the evaluative component. Without at least an imaginative understanding of what sort of evaluation is carried by words such as "coward," "cruel," or "person," these concepts would simply be arbitrary ways of dividing up the world.[6]

Attitudes and Souls

In the *Philosophical Investigations*, Wittgenstein writes, "My attitude towards him is an attitude towards a soul. I am not of the opinion that he has a soul" (PI § 178). I want to suggest that to treat a severely neurologically damaged child as a person—or, in Wittgenstein's more apt phrase, as a "soul"—involves taking up a certain attitude toward him. By "attitude," I mean the kind of stance implicit in our dealings with other persons, such as the recognition that a person deserves a special kind of respect, that he or she is to be given a proper name (rather than, say, a number), that he or she is to be referred to with the pronouns "he" or "she" and "him" or "her" rather than "it," and so on. Taking up this attitude toward a person does not follow necessarily from any fact about him; it is not a logical consequence of anything that I believe about him ("I am not of the opinion that he has a soul"). Rather, the attitude—an attitude toward a soul—is built right into the language that we use to describe him.

Our uncertainty about severely damaged children, it seems fair to say, is less about whether they are properly described as "persons" than about what attitude toward them it is appropriate to adopt and what language best expresses this attitude. The problematic ethical questions that bioethicists tend to ask about such children—can we allow this child to succumb to an easily treatable pneumonia? are we obligated to ventilate a vegetative patient indefinitely at the parents' request? can we transplant a heart from an anencephalic child?—are problematic exactly because they are situated within this broader uncertainty. If we are uncertain how to answer these ethical questions, it is because our broader, more general attitudes toward these children are themselves ambivalent and poorly articulated.

Now, to take up this sort of attitude toward a human being—an attitude toward a soul—is in part to recognize that she is the proper object of certain moral attitudes. That is, it is to recognize that we have duties toward her, that she has rights, and so on. But an attitude toward a soul, I want to suggest, is not solely a moral attitude. It encompasses all the complex ways that we treat our fellow human beings. Some of these may involve moral attitudes, such as the idea that all human lives have dignity, or that all lives are to be valued equally. But others seem closer to matters of custom, manners, or tradition, such as the idea that a person should be referred to by a name, or that when a person dies she should be given a funeral. Our moral attitudes are situated within this much broader family of attitudes.

All this is to say that our ambivalence about how to behave toward these severely damaged children is not solely a moral ambivalence. It is a broader ambivalence about what attitude is appropriate to such damaged, limited human lives. Let me give an example. I would find it very disturbing, though not in any strictly moral sense, to see parents hold a birthday celebration for an anencephalic child—to bring him a cake, put a birthday hat on him, sing to him, and so on. Why? Partly, perhaps, because the child will never be capable of recognizing the significance of the event; partly also because the celebration would remind me of the gap between an anencephalic and ordinary children, like my own. But I think these are part of a larger reason, which is that a birthday celebration implies that an anencephalic is a child like any other. And celebrating the birth of an anencephalic suggests that we take up the same attitudes towards *her* that we take toward other children: that this child will be a part of the family like any other child, that her life will have a narrative like that of an ordinary human being, from birth through childhood and adulthood to death. It suggests that this is the kind of being for which a birthday celebration is appropriate. And what seems painfully obvious is that an anencephalic is not such a being; that the reasons why we celebrate a birthday are absent here; that the passage of another year of life can have no meaning for a being without a cortex.

Now, by saying that taking up such attitudes would not be morally disturbing, I do not mean to trivialize them or to imply that they would have no moral resonance. I only want to make the point that these attitudes should not be reduced to their moral components; that the word "moral" does not fully capture what it is to take up such an attitude; and that when we take up the wrong attitude, it does not seem quite right simply to call the attitude immoral. I might say, for example, that it is immoral for parents to refuse to give their children names, or, even worse, to give them names like, say, Fluffy, or Rover; it is immoral, but it is not *simply* immoral. It seems closer to the truth to call it unsettling, or even just creepy, because it seems to represent an inappropriate attitude to take up toward a child. It suggests the attitude one takes toward a thing or a pet; it denies the child her humanity. Many of us would find it deeply disturbing. But the reason is the reverse of the reason the birthday celebration for an anencephalic is disturbing, which is that the celebration seems to attribute to that child a humanity that she does not have.

This may help to explain some of the rancor and division over the issue of using anencephalics as organ sources. Opposition has little if anything to do with cruelty, suffering, or violations of rights, but rather is often expressed in terms like "dignity" and "respect." This kind of language expresses a discomfort with the attitude toward the anencephalic that using them as organ sources seems to represent. It comes close to treating them as objects or things. But what is wrong with treating them as objects, one might say? They are never-conscious beings, more like corpses than human beings. The answer, I suspect, has something to do with what Wittgenstein is gesturing toward when he writes, "The human body is the best picture of the human soul" (PI § 178). Anencephalics are, after all, living infants who often look very much like ordinary infants. It should not be all that surprising that many people (especially parents) find it difficult to take up the same attitude toward them that we do towards objects or even corpses.

Yet neither is it easy to take up the same attitudes toward them as toward ordinary human beings, and it should not be surprising that the attitudes of doctors are often much different from those of parents, or even the lay public. When I was a third-year medical student, my rotation in internal medicine took place at the county hospital in Charleston, South Carolina. I can remember following one of the interns on ward rounds, a particularly sharp and easygoing man who, when the rotation began, took the time to introduce me to each patient on the ward, most of whom were (this being a county hospital in South Carolina) poor and black. We passed through the room of one patient, an elderly woman who was, if not permanently vegetative, very close to it. She was getting no treatment other than tube feedings and hydration. The intern's instructions to me were roughly this: "Think of it this way. She's a plant; you're the gardener; your job is to make sure she is watered." And then we moved on to the next patient.

To understand why this kind of remark is in equal parts callous, deeply embarrassing, and, in a despairing way, weirdly appropriate, you probably will need to have spent some time in a county (or Veterans Administration) hospital, where exhausted and often bitter residents take care of America's sick and disabled poor. This is the vacuum in which attitudes toward severely impaired patients develop, and the intern's remark reflects those attitudes: hostility at having to take care of such a patient, a sense of futility surrounding her future, and a sensibility trained to ignore deeper questions surrounding life and death.

Forms of Life

Another way of making these points would be to say that we have developed a certain kind of language that we use in describing ordinary persons, their behavior and mental lives, and our behavior toward them. It includes not only moral language, but also the language of religion, kinship, ritual, politics, and so on. The moral question is whether, and in what ways, this language extends to such severely damaged children.

But putting it this way makes things sound simpler than they are. It sounds as if the attitudes we take up toward other beings are essentially a matter of what we *decide*. But this is not quite right. Our attitudes toward other beings are rooted in, if not exactly instinctive behavior, still more or less thought-free behavior—reactions rather than conscious deliberations. (This is part of why Wittgenstein contrasts attitudes with opinions.) Now, I do not mean to suggest that all of our dealings with other persons are mechanical and without thought—that we do not make conscious decisions to argue with other people, or to flirt, or to joke with them. What is does mean is that in the background of all this behavior is the attitude that this is *the kind of being* with which you can argue and flirt and joke. And this attitude is not something that we ordinarily decide upon; it is simply what we do. As Wittgenstein says, "The essence of the language game is a practical method (a way of acting)—not speculation, not chatter" (PO, p. 399).

Now, does this mean that what attitude we take toward another being is something over which we have no control, or that it cannot be influenced by the will? Of course not. To take a superficial example: think, for example, of the different attitudes a doctor takes toward a human being, first, when he is percussing his chest, and next, when he is playing against him on the basketball court. An attitude toward a human being as the object of a diagnostic examination is different from an attitude toward a human being who is about to take you to the hoop. Patients are different from opponents. And which attitude you take up is more a kind of unconscious reaction, depending on the context, than a conscious decision. It is less like changing your mind than like falling back into a habit.

Another way to put this point is to say that our attitudes toward other beings are built into the language that we use to describe them, and the language is embedded in a way of behaving toward them—what Wittgenstein calls a "practical method." This practical method is not something that is best described as deliberative action, but something that is reactive and habitual. As Wittgenstein puts it: "The origin and the primitive form of the language game is a reaction; only from this can more complicated forms develop. . . . Language—I want to say—is a refinement. 'In the beginning was the deed'" (po, p. 395). This point is connected to another problem in the debate over personhood, or the quest to find some key difference between these severely damaged children and a person and to put your money on that difference as the morally crucial characteristic. Philosophers have tried on various occasions to name, as the morally crucial characteristic, consciousness, the capacity for speech, and the capacity to feel pain, among many others. The point I would like to make, following Cora Diamond, is that it is not enough just to ask whether a given characteristic is morally important; we also have to ask what a particular group of human beings has *made* of that characteristic.[7] A biological characteristic becomes something for moral consideration when human beings make something of that characteristic: in their religion, art, literature, rituals, institutions—and in their ethics. In some cultures a young girl's menarche is of tremendous moral significance, while in others it is not. Some cultures make a lot of the differences between men and women, while others do not. Menopause is important for some cultures, for others not. The birth of twins may be, depending on the culture, a sign from God, a curse, or nothing other than a reason to dress the children alike and give them rhyming names. My point is that saying something about the moral significance of these biological characteristics (or lack thereof) is not just a matter of saying something about those characteristics themselves, but of the form of life in which those capacities do or do not make a difference.

Think, for example, of the ways that we North Americans distinguish between the concepts of *pets*, *livestock*, and *vermin*. In the case of rabbits, there are no biological differences that distinguish between the categories. We eat rabbits as livestock, we keep them as pets, and we poison them as vermin when they get into the garden. Our attitudes toward rabbits differ dramatically in all three cases, and so does the way we treat them morally, but these differences do not depend on their biological characteristics. Another example: Westerners, unlike many Asians, find it horrifying and repulsive to think about eating a dog. Is this explainable in terms of the dog's characteristics? No, or at least not wholly; and if you want to understand why a people find this horrifying, you have to understand what a culture has made of these characteristics.

This helps explain why, in the case of neurologically damaged children, constructions such as "best interests" often seem less than helpful. We can say that severely damaged children have an interest in avoiding pain and in things that give them pleasure, but we cannot say much more. It is even hard to say, sticking solely to the language of interests, why severely damaged children like this have an interest in avoiding things that many people have a gut reaction against, like being used as a living organ donor, or being anesthetized and used as teaching instruments, or, to use an example from a science fiction short story, having their skin used as handbag leather after they die. If we want to say why we find such things repellent and horrifying, we usually fall back on concepts such as harms to dignity or the "symbolic value" of a body, which may not explain much but at least get us closer to the idea that these actions represent objectionable attitudes toward such children.

Conclusion

I have tried to express some of my misgivings about the notion that we can decide how to behave toward neurologically damaged children based solely on their capacities, or even by asking what is in their interests. How we think about and behave toward them is tied to the attitude we take toward them, which is in turn tied to a form of life. But what does this tell us?

First, it suggests that there is no single morally correct attitude to take toward such infants, but rather a range of attitudes, which are in turn embedded in particular cultures. It would not surprise me, for example, to hear an anthropologist speak about one culture that revered such damaged children and another culture that simply discarded them, and that each attitude was tied in complex and subtle ways to the culture's religion, structures of kinship, beliefs about health and illness, and so on. Even in our own (Western) culture(s) we hear a broad range of opinions on the appropriate behavior toward such children that touch on everything from infanticide to disability rights to the sacredness of every human life. The capacities and interests of such damaged children do not give us, or any culture, a determinate answer on how to behave toward them. And how we do behave toward them cannot be thought about separately from all of our other cultural resources.

Second, I do not believe we can completely separate how we think about damaged children from the way we think about ordinary children. A culture might well think about severely damaged children in very different ways depending on, say, whether they think of children as a kind of family property, or whether they think of a damaged child as a kind of curse or divine retribution, or whether they think of the deaths of a number of small children as natural or unavoidable.[8] In such contexts one can imagine the death even of a less severely impaired child to be a matter of indifference, or even something to be desired. But these are not the ways we in industrialized Western countries have come to think of children. We have inherited a certain ideal of the family and its importance, and while our attitudes toward children are complex and often contradictory, it is certainly true that we devote considerable resources to thinking about the rights of children and our duties towards them.[9] We treat them as, if not ends in themselves, at least ends to be, and many people think of their children as the most precious and important things in their lives. Whatever attitudes we hold toward severely neurologically damaged children will have to be reconciled with these more general attitudes toward children and family life.

Finally, and perhaps most crucially, whatever our attitudes toward severely impaired children, they will have to be reconciled with broader cultural understandings about the purpose and significance of human life. By this I mean not just what Wittgenstein calls *Lebensformen*, or forms of life, but that dimension of *Lebensformen* relating to questions such as what constitutes a meaningful life, or when a life has sense, or what kind of life counts as a success or a failure. Different cultures and different eras have asked and answered such questions in dramatically different ways, of course, and many individuals may answer them differently even within a single culture, especially in immigrant countries such as the United States and Canada. But understandings about questions such as these form the backdrop against which human practices take place and help shape our conceptions of the moral dimensions of those practices, including our actions regarding severely impaired children.

What is particularly difficult to reconcile here are our attitudes toward such children and certain widely shared Western views about the meaning of human life. I have in mind the cluster of convictions surrounding what Charles Taylor calls "the affirmation of ordinary life." [10] These convictions locate significance in things like our families and the people we love, but also in meaningful work—the satisfaction of artistic or literary creativity, the sense of higher mission involved in social or political activism, the gratification of doing a job well, the fulfillment of a moral duty to provide for one's family and loved ones. These kinds of convictions locate significance largely in the individual and how he or she chooses to live a life. Thus, the significance of a life is intimately tied to the choices a person makes, which may involve, for example, developing a relationship with God, fulfilling a calling, carrying out one's duties, or many other things.

What makes these understandings about the significance of life difficult to reconcile with our attitudes toward children is that such understandings make a meaningful life inaccessible to any child with severe neurological impairment. (And, for that matter, to many children and adults with less serious damage.) If a person will never be capable of appreciating the emotional bonds of family, will never be able to find meaning through the life of work, and will never be able to return love to another human being, then she will not be able to live the kind of life to which many Western cultures give meaning. This is not the only way of seeing the lives of such children, of course. Things might be otherwise in cultures where meaning is bound up with a being's place in the natural order, or with the transmigration of souls, or any number of other cosmologies. But these are not the cosmologies that form the selves of most Westerners. Perhaps this is why the lives of profoundly damaged children strike many Westerners as especially tragic. These children throw into vivid relief the contrast between the kind of life that allows us to achieve the goods that make life worth living and the very different kind of life that lies ahead for such a child. [11]

This may help us to understand the internal contradictions of clinical decisions for these children and the reasons why they often appear so intractable. On the one hand, we are understandably wary of withholding or withdrawing beneficial treatment for such children. We are the inheritors of a strong tradition of rights and equality that makes us reluctant to withhold treatment from a person on the grounds of her intelligence or disabilities. We also see that the lives of these children may have deep significance for their families. Yet on the other hand, we recognize that these lives fail to meet the criteria by which we count our own lives as meaningful. We try to convince ourselves that we should protect vulnerable lives, but we cannot imagine this as a life we would want to continue living. We say all lives deserve respect, but our measure of the good life for ourselves does not include a life like this. We say that all lives are equal in the eyes of God, but we wonder why God has allowed such a life to come into being.

Notes

1 John Arras, "Toward an Ethic of Ambiguity," *Hastings Center Report* 14 (Apr. 1984): 25–33.

2 For a sampling of such writings on personhood, see H. Tristram Englehardt, "Ethical Issues in Aiding the Death of Young Children," in *Intervention and Reflection: Basic Issues in Medical Ethics*, 4th ed., ed. Ronald Munson (Belmont, Calif.: Wadsworth, 1992), pp. 119–26; Joseph Fletcher, "The Cognitive Criterion of Personhood," *Hastings Center Report* 4 (Dec. 1975): 4–7; Mary Anne Warren, "The Moral Significance of Birth," in *Feminist Perspectives in Medical Ethics*, ed. Helen Bequaert Holmes and Laura M. Purdy (Bloomington: Indiana University Press, 1992), pp. 198–215; Michael Tooley, "Abortion and Infanticide," *Philosophy and Public Affairs* 2, no. 1 (1975): 29–65; John A. Robertson, "Involuntary Euthanasia of Defective Newborns," *Stanford Law Review* 27 (1975): 246–61; Daniel Dennett, "Conditions of Personhood," in *The Identities of Persons*, ed. Amélie Oksenberg Rorty (Berkeley: University of California Press, 1976; and various essays collected in *What Is a Person?* ed. Michael F. Goodman (Clifton, N.J.: Humana Press, 1988).

3 Englehardt, "Ethical Issues in Aiding the Death of Young Children," p. 120.

4 Bernard Williams, *Ethics and the Limits of Philosophy* (Cambridge, Mass.: Harvard University Press, 1985), p. 129.

5 Cora Diamond has made this point more forcefully than I have here in her extraordinary essay "Eating Meat and Eating People," in *The Realistic Spirit: Wittgenstein, Philosophy, and the Mind* (Cambridge, Mass.: MIT Press, 1995), pp. 319–34.

6 Williams, *Ethics and the Limits of Philosophy*, p. 141.

7 See Diamond, "Eating Meat and Eating People."

8 For two very different cultural worldviews and their relationship to children, see, e.g., Nancy Scheper-Hughes, *Death without Weeping: The Violence of Everyday Life in Brazil* (Berkeley: University of California Press, 1992) and Anne Fadiman, *The Spirit Catches You and You Fall Down: A Hmong Child, Her American Doctors, and the Collision of Two Cultures* (New York: Farrar, Straus and Giroux, 1997).

9 For a provocative and deeply moving account of some of these contradictions, see John D. Lantos, *Do We Still Need Doctors?* (New York: Routledge, 1997).

10 See Charles Taylor, *Sources of the Self: The Making of the Modern Identity* (Cambridge, Mass.: Harvard University Press, 1989).

11 One of the few works I know of that takes on these deeper questions about dying children is the wonderful essay by Margaret Mohrmann, "Are Children Our Future? Reflections on Destiny and Dying Children," presented at a conference on Bioethics and Human Destiny: Jewish and Christian Perspectives, Loma Linda Center for Christian Bioethics, Loma Linda, California, February 1997.

7 Why Wittgenstein's Philosophy Should Not Prevent Us from Taking Animals Seriously
David DeGrazia

In recent years, academic philosophers and the broader public alike have given considerable attention to specific ethical issues concerning our treatment of animals and to more fundamental questions about animals' moral status—in short, to animal ethics. More and more philosophers working in this area conclude that animals have significant moral status and that many or most of the prevailing practices and institutions of animal use at least in anything resembling their current forms, are morally indefensible.[1]

There are those who believe, however, that one or more elements in the philosophy of Ludwig Wittgenstein weigh against taking animals quite so seriously. First, some scholars argue that Wittgenstein's philosophy of mind and of language justify skepticism about animals' mental life.[2] Such skepticism is important, because the assignment of significant moral status to animals is commonly thought to depend on the thesis that they suffer (on a sentience critierion), have desires and beliefs (on an agency criterion), possess a minimal self-awareness or sense of time (which may be necessary for suffering), or the like.

A second possible roadblock in the argumentative path to an animal protection view is Wittgenstein's antitheoretical conception of ethics. To reject the status quo regarding animals means rejecting commonly held, fundamental moral assumptions, but only extensive argument and perhaps some manner of ethical theorizing would seem to justify a general expectation that people do so. Yet to call for such a radical change in attitudes and practices, with philosophy leading the charge, seemingly runs counter to Wittgenstein's dictum that philosophy "leaves everything as it is."[3]

The thesis of this essay is that Wittgenstein's philosophy should not prevent us from taking animals seriously. Regarding the first line of argument described above, Wittgenstein's views do *not* have highly skeptical implications about animal minds. As for the second line of argument, while Wittgenstein's views apparently support a strongly antitheoretical stance in ethics, such a stance, properly understood, does not preclude the justification of major moral reform.

Wittgenstein on the Mental Life of Animals

Some scholars take Wittgenstein, or at least aspects of his philosophy, to suggest that animals are incapable of thought, beliefs, desires, suffering, or the like. Frey, for example, invokes Wittgenstein's philosophy of mind and language in arguing that animals are not self-aware, a thesis that serves as a premise in an argument that animals lack desires and interests—and therefore significant moral standing.[4] Although he disagrees with Wittgenstein, Rollin takes his views to imply that animals lack thought, which is deemed necessary to get one beyond the particularities of sensation; the upshot is that animals are "stuck in the present" and therefore incapable of suffering (since suffering, unlike pain, is argued to involve a sense of time).[5] Leahy, to whom it is hard to ascribe a precise line of argument, draws extensively from Wittgenstein and somewhat similarly contends that "[i]t is as primitive beings [*sic*] that we must assess the claims of proper treatment [for] animals"; this is the basis on which he opposes the animal protection movement.[6] Singer captures the vague sense that Wittgenstein's views may pose a threat to significant moral status for animals when he writes that "there is a hazy line of philosophical thought, deriving perhaps from some doctrines associated with . . . Wittgenstein, which maintains that we cannot meaningfully attribute states of consciousness to beings without language."[7]

This section begins with an argument to a skeptical conclusion, examines Wittgenstein's statements about animals' mental life, and finally attempts to bring the evidence into reflective equilibrium with a nonskeptical interpretation. Let us begin with a line of reasoning, which might be naturally ascribed to Wittgenstein, that concludes that animals cannot think. For reasons of space, rather than examining parallel arguments about other kinds of mental states and capacities, let us use thought as a representative of states that one might believe Wittgenstein denies to animals—including propositional attitudes such as beliefs and desires, and states such as suffering that might seem to require a degree of self-awareness or temporality.

1. Animals lack language.
2. Thought requires language. Ergo,
3. Animals lack thought.

The conclusion clearly follows from the premises. But do we have good reason to attribute the premises to Wittgenstein?

With one qualification, to be explained in a moment, premise 1 is reasonably ascribed to Wittgenstein. Consider first what he terms a "private language"—a hypothetical language whose words refer to putative private objects (e.g., what some philosophers call "qualia"), which are inexpressible in principle to anyone other than the sole possessor of the language. In his famous private language argument, Wittgenstein argues that the very idea of such a language, even in the case of humans, is incoherent (PI §§ 243ff.). There is no need to review that argument here, because a private language in the sense defined is the last thing one could reasonably attribute to animals.[8] Surely an animal language would have terms or signs that referred to things in the world, such as food or water. But, as suggested by the textual evidence reviewed below, Wittgenstein held (with a qualification) that animals lacked that sort of language as well. Genuine language, he thought, involves linguistic rules as well as public practices that permit correction and confirmation of language users; but animals do not engage in practices of the requisite complexity. My contention is not that Wittgenstein ruled out such animal practices a priori—in which case no nonhuman-ape language study, no matter how successful, could make the grade—but that he did not think animals, in fact, displayed such achievements.

Thus Wittgenstein stated that animals "do not use language—if we except the most primitive forms of language" (PI § 25). The qualification, then, is that Wittgenstein seemed open to the possibility that animals have language of "the most primitive forms." Since the bulk of his investigations of language stress the requirements of rule following and practices of correction and confirmation, and since Wittgenstein apparently did not believe animals meet those requirements, the most liberal thesis we can attribute to him is that some animals might display "proto-language," or language in some very limited sense. The language of most interest to him was the set of complex linguistic practices humans participate in, which are intertwined with elaborate forms of social life and thought.

For this reason and for the sake of argument, let us set aside the qualification and grant that Wittgenstein held that animals lack language, as premise 1 asserts. Did Wittgenstein hold premise 2, that thought requires language? Perhaps the strongest evidence that he did is his attack on the Augustinian conception of language learning. On this view, children are actively thinking about the world before learning language, at which time they associate conventional words with full-blown thoughts they have already had: "Augustine describes the learning of human language as if the child came into a strange country and did not understand the language of the country; that is, as if it already had a language, only not this one. Or, again: as if the child could already *think*, only not yet speak. And 'think' would here mean something like 'talk to itself'" (PI § 32). Against this view, Wittgenstein argues that signs, which would include images or other mental media, are by themselves semantically inert, not self-explicating; a sign's meaning is determined by how it is used in practices that maintain standards for correct use.[9] This appears to support the idea that, for Wittgenstein, thought without language and associated forms of life (complex patterns of activity) is impossible, a point that extends to infants as well as animals. Without language, the argument goes, thought would presumably require mental media intrinsically endowed with meaning, an inconceivable hypothesis on Wittgenstein's view.

Yet what Wittgenstein actually said about animals does not easily square with this conclusion.[10] He states in the *Philosophical Investigations* that dogs cannot simulate pain (PI § 250), suggesting that they can have pain. The remark that "we do not say that *possibly* a dog talks to itself" (PI § 357) apparently precludes a canine internal monologue that never gets expressed publicly. The natural expression of an intention is visible in "a cat when it stalks a bird; or a beast when it wants to escape" (PI § 647). A dog can be afraid his master will beat him, but not that his master will beat him tomorrow (PI § 650). We can imagine an animal angry, frightened, happy, unhappy, startled—but not hopeful (PI p. 174). "A dog believes his master is at the door," but not that his master will come the day after tomorrow (PI p. 174). Somewhat cryptically, he remarks that "if a lion could talk, we could not [or: might not be able to] understand him" (PI p. 223).[11] Also, a "dog cannot be a hypocrite, but neither can he be sincere" (PI p. 229). Here is a clue to interpreting some of these passages: "Can only those hope who can talk? Only those who have mastered the use of a language. That is to say, the phenomena of hope are modes of this complicated form of life" (PI p. 174).

In *Zettel*, one reads that "dogs feel fear, but not remorse" (z § 518). "Only someone who can reflect on the past can repent" (z § 519). Here is another general clue: certain concepts apply only to those who possess a language (z § 520). We cannot say that a dog means something by wagging its tail (z § 521). Moreover, we "should hardly ask if the crocodile means something when it comes at a man with open jaws. And we should declare that since the crocodile cannot think there is really no question of meaning here" (z § 522). "There might be a concept of fear that had application to beasts, and hence only through observation. . . . The verb that would roughly correspond to the word 'to fear' would then have no first person and none of its forms would be an expression of fear" (z § 524). On the basis of behavior under certain circumstances, we can say that someone is sad—a point that applies to dogs as well as persons (z § 526).

Tending to confirm such statements are passages from *Remarks on the Philosophy of Psychology*. For example, Wittgenstein suggests that an orangutan can be angry but not hopeful (RPP vol. 1, § 314) and that crocodiles do not hope (RPP vol. 2, § 16). Less redundantly, he also suggests that it is uncertain whether lower animals, such as flies, feel pain (RPP vol. 2, § 659).

Let us take these passages together. (Absent a special reason to think otherwise, such as the remark just noted about lower animals, let us assume that what Wittgenstein says of one species of animal, he means to apply to other species.) Wittgenstein has asserted—in each case either directly or in his indirect, ironic way—that animals can have pain; be afraid (though in one place he speaks of a specialized concept for application to animals), as well as angry, happy, unhappy, sad, or startled; want something; express an intention; and even believe something. On the other hand, Wittgenstein denies that animals can talk to themselves, have fears or beliefs about tomorrow or the day after, hope, feel remorse or regret and reflect on the past, demonstrate hypocrisy or sincerity, mean something by an action or gesture, or think. And if they could speak, we would (or might) be unable to understand them. At the same time, some "lower" animals may not even be able to experience pain. Do these claims form a coherent body? Are they compatible with the skeptical argument sketched earlier?

Let us begin with the most easily handled mental states and work up to the hardest. First, Wittgenstein ascribes sensations to many animals. A few passages from *Zettel* shed some light on the theoretical basis: "The concept of pain is characterized by its particular function in our life" (z § 532); "Pain has *this* position in our life; has *these* connections" (z § 533). These passages tersely convey Wittgenstein's view of sensations as states of an individual that are typically expressed by characteristic behaviors in certain recognizable circumstances and within certain forms of life. That pain is typically expressed by certain kinds of behavior is an essential part of the concept. The link to animals is that creatures whose behavior, circumstances, and forms of life suggest pain, do have pain, and similarly with other sensations. (Very primitive creatures might not qualify because even if their behavior is painlike, their forms of life might be too rudimentary.)

Wittgenstein also ascribes a considerable variety of emotional states to animals, such as fear, anger, and sadness. He provides a bridge between sensations and certain emotions: "Only surrounded by certain normal manifestations of life, is there such a thing as the expression of pain. Only surrounded by an even more far-reaching particular manifestation of life, such a thing as the expression of sorrow or affection. And so on" (z § 534). The behaviors—more broadly, the forms of life—necessary to express emotions may be somewhat more complex than those expressive of sensations, but the behavior of dogs and many other "higher" animals invites the ascription of certain emotions.

But within what limits? Wittgenstein bars the attribution to animals of emotions (as well as cognitive states and character traits) that require a complexity of cognition available only to language users. Without language and a sufficiently complex form of life, he figures, it would be impossible to form the thought of the day after tomorrow, to reflect in depth about the moral quality of one's actions, to talk to oneself. Nor, without language, could one be sincere or hypocritical, mean something by a gesture or action, or hope. One implication is that language is necessary for mental states referring to distant points in time, suggesting that any sense of time possessed by animals is quite limited. As with sensations, then, Wittgenstein grants to animals more-complex mental states that are vindicated by the appropriate behaviors, circumstances, and forms of life. Presumably, this is how he regards acting with an intention and the corresponding desire, as with a cat stalking a bird.

But do not intentions require beliefs, such as "there is something ahead to be stalked"? And do not various emotions require beliefs? If the dog fears his master will beat him, must he not believe that his master might do so? Moreover, Wittgenstein explicitly states that a dog can believe his master is at the door. But notice: to believe that *p*, one must *think* that *p*. Belief entails thought (dispositional, if not occurrent). Yet above I spelled out a seemingly strong case that Wittgenstein's philosophy precludes animal thought, a thesis that appears to be vindicated by the crocodile passage. How are we to understand Wittgenstein on animal thought?

Kenny's interpretation, which is very much in line with what has just been argued, strikes me as plausible: "There are thoughts which only a language-user can have, as well as thoughts which animals can share: a dog can believe that his master is at the door, but not that his master will come the day after tomorrow, because he cannot master the complicated language in which alone such a hope can be expressed." [12] *A languageless being can think that* p *only if the thought* p *is expressible by her nonlinguistic behavior.* A dog shows that she believes her owner is at the door if, at the jingling of keys, she rushes to the door, showers affection on her owner if it is he, acts surprised or suspicious if it is someone else, and so on. Nothing a dog could do (science fiction aside) could show that she believes that her owner will be showing up on the Fourth of July or that Satan exists—thoughts for which it is plausible to require the conceptual rocket of language.

But if Wittgenstein accommodates simple animal thoughts, why does he say that "since the crocodile cannot think there is really no question of meaning here"? If thought is ruled out, so is meaning; so let us focus on the former: why cannot a crocodile think "there's food," nonlinguistically? Is Wittgenstein distinguishing between the differing capacities of different animals: dogs, but not crocodiles, can have simple thoughts? That is possible, since reptiles are more primitive than mammals and exhibit less nuanced behavior; Wittgenstein asserts that we ascribe mental states only to what behaves like a human being (PI 281), and mammalian behavior is more humanlike than reptilian behavior. Another possibility is that by "think," Wittgenstein here means the higher-level sort of thought that he ruled out for nonspeaking infants in arguing against the Augustinian view when he said, " 'Think' would here mean something like 'talk to itself' [conduct an internal monologue]." It is, after all, more robust, human-typical sorts of thinking that most interest Wittgenstein throughout his work. And he holds that languages are built upon more primitive reactions and routines, such as crying when in pain and tending to others when we see that they are in pain;[13] perhaps languageless thought is another such primitive beginning.[14] A third possibility is that his crocodile remark simply represents an inconsistency with his overall view of animals, thought, and language. *Zettel* is a far less edited work than the *Investigations* and the Wittgenstein corpus contains much that was experimental.

What about the lion?[15] Wittgenstein's remark emphasizes that there must be sufficient overlap in forms of life for one life-form, say, humans, to understand another, say, lions. Meanings within a language are grounded in shared forms of life; successful explanations of meaning depend on common dispositions to react in certain ways to particular signs (as explained in the passages on rule following). But if the remark's correct translation is indeed that we *could* not understand the lion, it seems overstated. If a lion could speak, we could probably understand much of what he had to say about food, danger, relaxation, and lust. But what is most unfortunate about the statement is that it invites the misreading that Wittgenstein thinks each lion has an internal monologue that he cannot manage to express publicly—talking only to himself.

Now we can reconcile the nonskeptical direction of Wittgenstein's remarks about animals with his philosophy of language and philosophy of mind, which seem to generate the skeptical conclusion drawn above. What Wittgenstein's philosophy suggests is not that animals as a class cannot think, but rather that language, which serves as the major vehicle of human thought, cannot serve as the vehicle of animal thought, because animals lack language. Animal thought is ascribable on the basis of behavior that (in context, and given their forms of life) expresses what we would normally take to be thought; the vehicles of thought might be mental images put to certain uses, so long as the correctness or incorrectness of thoughts could be checked by interactions with the physical world.[16] (For example, a dog could check the thought that a smelly shoe is edible—a thought that might be carried by some vague, multisensory representation of eating the shoe—by trying to eat it.) It turns out, then, that Wittgenstein is not skeptical about animals' mental life. Indeed, his attributions to animals are mostly commonsensical.[17]

Wittgenstein, Anti-Theory, and Moral Reform

Wittgenstein does not deny that animals have the sort of mental life that arguably underlies significant moral status. Does he suggest that the justification of major moral reform (for example, by appeal to an ethical theory) is not the proper business of philosophy? If so, then for those sympathetic to Wittgenstein's views, it would appear that philosophy cannot justify a call for major change in our animal-using practices. An even more reactionary thesis that might be attributed to Wittgenstein is that major moral reform cannot be justified at all (whether by philosophical work or anything else).[18] I begin this section by acknowledging a good case that Wittgenstein's philosophy supports anti-theory in ethics, before arguing that the latter does not preclude the justification of major moral reform. I leave open, however, whether such justification should be regarded as philosophical work.

The following passages on the proper role of philosophy, all from the *Investigations*, illustrate the textual basis for attributing anti-theory in ethics to Wittgenstein.[19] "[W]e may not advance any kind of theory. . . . We must do away with all *explanation*, and description alone must take its place" (PI § 109). "Philosophy simply puts everything before us, and neither explains nor deduces anything" (PI § 126). "In philosophy we do not draw conclusions. . . . Philosophy only states what everyone admits" (PI § 599). "If one tried to advance *theses* in philosophy, it would never be possible to debate them, because everyone would agree to them" (PI § 128). "Philosophy may in no way interfere with the actual use of language; it can in the end only describe it. . . . It leaves everything as it is" (PI § 124). Readers familiar with Wittgenstein's work will recognize these statements not as describing philosophy as it is actually practiced, but as prescribing how it should be practiced. (Note that these very statements—assuming they count as philosophy—violate their own injunction to limit philosophy to description.)

Clearly, much work by animal protection philosophers does not conform to the role Wittgenstein urges for philosophy. Ethical theories and more specific theories of the moral status of animals are advanced, along with myriad particular theses with which not everyone (or even a majority) agrees. Philosophical arguments in support of an animal protection view hardly leave "everything as it is."

But does Wittgenstein's philosophy really support such a reactionary attitude about ethics? Maybe we do his views an injustice by reading off so directly from his most theory-hostile remarks what he might say about moral reform or even ethical theory. Let us review, then, the general considerations that stand behind these remarks.

Wittgenstein believes that human thought—whether philosophical, theological, or any other kind—cannot achieve a perfectly privileged "God's-eye view" from which to evaluate claims of truth or moral rightness, suggesting that the traditional philosophical quest for such a radically objective standpoint is misconceived. We are confined, Wittgenstein thinks, to the everyday practices in which human discussion, reflection, and criticism take place. Any evaluation or critical thinking occurs within some practice and is subject to its standards; human practices are ultimately made possible by shared dispositions or natural reactions that are not themselves rationally grounded,[20] hence the inability of reason to soar beyond the contingencies of human life to a "view from nowhere."

This position, if correct, may well discredit ethical rationalism, according to which certain nonmoral considerations or facts make a particular ethical view rationally necessary. Kant was therefore wrong to think that pure reason could produce, for all possible rational beings, a compelling reason to be moral (namely, that immorality involves a kind of contradiction) and a standard for determining right action (the categorical imperative). But does the reasoning of the previous paragraph imply that there is no place for ethical theorizing or for moral reform inspired by philosophical reflection?

Answering affirmatively, one might argue that there is no authoritative framework, such as a religion or a particular moral outlook, from which substantive ethical disagreements can be independently resolved.[21] We take part in moral practices, but we cannot jump entirely outside of them and decide which moral practice, framework, or tradition is authoritatively best. In response to this view, from the fact that all ethical reflection takes place within a practice it hardly follows that one cannot (1) correctly or incorrectly judge among certain options considered within a practice (say, that of moral discourse itself), (2) develop a theory that accounts well for reasoning within a practice and then employ the theory in novel or difficult areas,[22] or even (3) correctly or incorrectly judge the superiority of one practice or tradition (say, liberal democracy) over another (say, apartheid) on the basis of standards—if there are any—that are accepted by both.[23]

Thus, the Wittgensteinian idea of entrenchment in practices does not so obviously support anti-theory in ethics and opposition to moral reform. But what this Wittgensteinian idea suggests and what Wittgenstein himself believed, or would believe, are not necessarily the same. The former is a philosophical view that can be fleshed out plausibly or implausibly; the latter is the conviction of a specific human being. Based in part on his statements about philosophy, it is highly probable that Wittgenstein himself would not regard ethical theory and the justification of reform as philosophical work; these passages also suggest that he was opposed to any sort of ethical theory. I believe that these attitudes are quite mistaken and that they are reflected in some of the most poorly argued, unimaginative, and dogmatic portions of the Wittgenstein corpus.

The only claim I will insist on here is important for those who accept Wittgenstein's anti-theory stance in ethics: this position does not preclude the justification of major moral reform (including reform of our animal-using practices). My claim is consistent with the Wittgensteinian thesis, with which I largely agree, that our reflection must take place within practices whose most basic standards and criteria we do not choose. For the practice of moral discourse (which includes ethical reasoning) contains critical resources that can be deployed in contesting controversial issues and defending moral reform.

But we must distinguish weaker and stronger versions of this thesis. The weaker version, which I think most Wittgensteinians would accept, is that the activities of challenging, disputing, defending moral views, and the like are intelligible and worthy of respect; one can justify moral reform in the sense that one's justificatory efforts represent a moral activity that is perfectly in order. (A positivist who regarded moral discourse as meaningless might disagree.) The stronger version, which many Wittgensteinians seem disposed to reject, is that certain positions on ethical issues can be shown superior to others and are thereby justified in a sense that implicates some notion of ethical objectivity (perhaps in the form of human intersubjectivity). Given many of his remarks about ethics, Wittgenstein himself would likely have resisted that stronger claim.[24] But our moral practices contain resources that make the idea of ethical objectivity perfectly intelligible.[25] Let me motivate this claim with a few programmatic remarks.

We come to participate in our moral practices with certain resources that are more or less taken for granted: elements of so-called common morality. But some of these materials may be used to question, criticize, and even reject other of those materials. In particular, our starting point of inherited practices includes not only (1) certain normative standards—such as equal consideration for humans and the idea that animals exist for human benefit—but also (2) certain critical tools, such as demands for consistency, satisfactory reason-giving, plausibility in the implications of one's claims, and fidelity to the facts. And there is no reason to assume a priori that all of common morality's materials of type (1) can survive the appropriate use of common morality's materials of type (2). After all, it is not always the case that widely accepted normative standards prevail because they hold up under sustained, critical scrutiny; other factors that may lead to and maintain widespread acceptance are self-interest, various forms of prejudice, socialization through propaganda and other means, laziness, and genuine misunderstanding.[26] Now the weeding out of untenable moral positions and the gradual movement toward more defensible ones can inspire, and be inspired by, the idea of objectivity in ethics.

Thus, rejecting ethical rationalism does not entail rejecting reasoned ethical criticism or calls for major reform—or even the claim that they can achieve objective justification (as asserted in the stronger version of the above thesis)—for these activities can be vindicated by use of the critical tools implicit in our shared moral practices. Admittedly, these programmatic remarks do not constitute a positive case for the claim that ethical positions can sometimes enjoy the support of objective justification. My aim here is simply to indicate why we should not infer from the entrenchment-in-practices idea that such justification is impossible, and to identify a line of argument that has the potential to legitimate such justification.

The broad model of ethical justification that I think best captures these points is the coherence model, also known as the model of reflective equilibrium,[27] but I will not explore it here—not least because to do so would be to leave Wittgenstein's philosophy far behind. Suffice it to say that we should not defer to common morality's embrace of the casual exploitation of animals just because we are convinced that Wittgenstein's philosophy supports a reactionary rejection of moral reform. Such reform can be justified by the judicious use of the critical tools embedded in our moral practices. And if Wittgenstein himself thought otherwise, he was wrong.

Notes

1 See, e.g., Peter Singer, *Animal Liberation*, 2d ed. (New York: New York Review of Books, 1990); Bernard E. Rollin, *Animal Rights and Human Morality*, 2d ed. (Buffalo, N.Y.: Prometheus, 1990); Tom Regan, *The Case for Animal Rights* (Berkeley: University of California Press, 1983); Mary Midgley, *Animals and Why They Matter* (Athens: University of Georgia Press, 1984); S. F. Sapontzis, *Morals, Reason, and Animals* (Philadelphia: Temple University Press, 1987); Rosemary Rodd, *Biology, Ethics, and Animals* (Oxford: Clarendon, 1990); Evelyn B. Pluhar, *Beyond Prejudice* (Durham, N.C.: Duke University Press, 1995); and David DeGrazia, *Taking Animals Seriously* (Cambridge: Cambridge University Press, 1996).

2 Some of these scholars are cited in the next section.

3 For works taking Wittgenstein to deny that theory can justify major practical changes, see, e.g., J. C. Nyiri, "Wittgenstein's Later Work in Relation to Conservatism," in *Wittgenstein and His Times*, ed. Brian McGuinness (Chicago: University of Chicago Press, 1982), pp. 44–68; Paul Johnston, *Wittgenstein and Moral Philosophy* (London: Routledge, 1989); and Michael P. T. Leahy, *Against Liberation* (London: Routledge, 1991), chaps. 7 and 8.

4 R. G. Frey, *Interests and Rights* (Oxford: Clarendon, 1980), pp. 101–10.

5 Bernard E. Rollin, *The Unheeded Cry* (Oxford: Oxford University Press, 1990), pp. 137–38, 140–43.

6 Leahy, *Against Liberation*, p. 166. See esp. chaps. 5 and 6.

7 Singer, *Animal Liberation*, p. 14.

8 Nevertheless, Frey, *Interests and Rights*, and Rollin, *The Unheeded Cry*, both take the private language argument to be a crucial part of a Wittgensteinian case for skepticism about animals' mental life.

9 See his extended discussion of rule following (PI §§ 143–242).

10 In the remainder of this section, I draw heavily from my "Wittgenstein and the Mental Life of Animals," *History of Philosophy Quarterly* 11, no. 1 (1994): 129–32, although some of my present remarks represent significant changes.

11 I added the parenthetical alternative translation. The German original of the second clause is "wir koennten ihn nicht verstehen." Cf. G. W. Levvis, "Why We Would Not Understand a Talking Lion," *Between the Species* 8, no. 3 (1992): 158.

12 Anthony Kenny, *Wittgenstein* (New York: Penguin, 1973), p. 150.

13 See, e.g., PI § 244; Z § 540–41; CV, p. 31 e.

14 In exploring this possibility, I have benefited from David Pears, "Have They Anything to Say? Wittgenstein's Views on Animals' Capacity for Thought" (unpublished paper), pp. 10–14.

15 For interesting discussions, see Levvis and also J. Churchill, "If A Lion Could Talk . . ." *Philosophical Investigations* 12 (4) (1989): 308–24.

16 Pears helped me see this point ("Have They Anything to Say?" pp. 12–13).

17 By contrast, Leahy, *Against Liberation*, takes Wittgenstein to be considerably more skeptical about animals' mental life. His interpretation is vitiated, however, by his frequent refusal to take Wittgenstein at his word (see, e.g., pp. 106, 121, 127, 135).

18 See Nyiri.

19 These passages are neatly collated in F. Ackerman, "Does Philosophy Only State What Everyone Admits? A Discussion of the Method of Wittgenstein's *Philosophical Investigations*," in *The Wittgenstein Legacy*, ed. Peter A. French, Theodore E. Uehling, Jr., and Howard K. Wettstein, Midwest Studies in Philosophy, vol. 17 (Notre Dame, Ind.: University of Notre Dame Press, 1992), pp. 246–47.

20 This is the point, I take it, of the mathematics example presented in PI § 185.

21 This seems to be the gist of Johnston's view.

22 I believe Gert employs this strategy in offering a descriptive account of moral reasoning in the form of a theory, in Bernard Gert, *Morality* (New York: Oxford University Press, 1988).

23 Naomi Scheman also recognizes that Wittgenstein's philosophy does not so clearly have reactionary implications, since it seems open to the possibility of significant change within a practice ("Forms of Life: Mapping the Rough Ground," in *The Cambridge Companion to Wittgenstein*, ed. Hans Sluga and David G. Stern [Cambridge: Cambridge University Press, 1996], pp. 383–410). However, she appears to recom-

mend replacing what initially seem to be philosophico-ethical problems with political agendas, whereas I leave open the possibility that ethical reasoning, criticism, and justification are properly part of the work of philosophy.

24 See Rush Rhees, "Wittgenstein on Language and Ritual," in *Wittgenstein and His Times*, ed. Brian McGuiness (Chicago: University of Chicago Press, 1982), pp. 99–101; and Johnston, *Wittgenstein and Moral Philosophy*, chap. 7.

25 Sabina Lovibond argues that Wittgenstein's philosophy supports moral realism— roughly, the view that moral statements can be literally true or false (*Realism and Imagination in Ethics* [Oxford: Basil Blackwell, 1983]). While I find Cora Diamond's criticisms of this interpretation pretty convincing ("Wittgenstein, Mathematics, and Ethics: Resisting the Attractions of Realism," in *The Cambridge Companion to Wittgenstein*, ed. Hans Sluga and Daniel J. Stern [Cambridge: Cambridge University Press, 1996], 226–60), Lovibond is surely right that certain features of moral discourse strongly suggest the idea of ethical objectivity.

26 Leahy seems to miss this point entirely. Thus, in considering ethical challenges to current practices of animal use, he appeals to those very practices in an effort to identify the appropriate ethical standards; for example, he thinks criteria for what suffering is "needless" are necessarily embedded in current uses of animals (*Against Liberation*, p. 198), implying that whatever we do *must* be right.

27 See, e.g., John Rawls, *A Theory of Justice* (Cambridge, Mass.: Harvard University Press, 1971), pp. 48–51; and Norman Daniels, *Justice and Justification* (Cambridge: Cambridge University Press, 1996). I defend my somewhat different version of the coherence model in *Taking Animals Seriously*, chap. 2.

8   Injustice and Animals
Cora Diamond

Reasons in ethics, Wittgenstein suggested, are like reasons in philosophy itself, or in aesthetics. They "draw your attention to a thing"; they "place things side by side"; sometimes they move things apart. Such reasons can change one's *Anschauungsweise*, one's way of viewing things.[1] This essay is about how the concept of injustice bears on our treatment of animals. It is about one moving apart of things and one placing of things side by side: the moving apart of justice and rights, and the placing side by side of ourselves and animals as beings to whom justice or injustice may be done.

I

If we ask about the bearing of the concept of injustice on our treatment of animals, we invite a familiar kind of approach. Justice is usually identified with respecting rights, and *in*justice, with violation of rights. Some people then argue that animals should be recognized as having rights, while their opponents claim that animals cannot have rights. And, indeed, the insistence that *rights* are crucial is what has given its name to the animal rights movement. Through the notion of rights, the movement distinguishes its own position from that of animal welfarism, which, roughly speaking, is the view that our practices using animals do not need to be fundamentally changed, but should be carried on without causing unnecessary suffering to animals.

Why, then, are justice and injustice usually tied so closely to rights? Two important background ideas encourage the assimilation of justice to respect for rights.

First, there is the idea that a right is something you can claim, you can demand it. You do not have to *beg* for it. And there is then a contrast with *charitable* treatment, or *generous* treatment, or *merciful* treatment, which you cannot demand as a matter of right but can only beg for. Frequently these are taken to be the two possible alternative responses if someone is being treated badly: the demand for rights or the request for mere kindness or mere charity.

The contrast between what we can demand as a matter of right and what we can merely beg for is then also frequently tied (in the case of justice and injustice to human beings) to ideas of dignity. Dignity is compatible with demanding one's rights but not, supposedly, with begging for charity or mercy. After all, the person from whom one demands one's rights has an obligation to grant what is thus demanded, but there is no obligation if mere charity is asked for, or generosity, or mercy. That point is sometimes also connected with the idea that kind or merciful treatment depends on the presence of compassionate feelings, while respect for rights is incumbent on us no matter what we may happen to feel.

The second important idea in the background when considerations of justice and injustice are tied closely to rights is that rights provide serious constraints on action. They protect fundamental interests that might otherwise be trampled upon whenever overall welfare might be improved by doing so.

It is that combination of ideas, then, that has made rights seem so important in contemporary thought about animals. If animals do not and cannot have rights, appeals that they be better treated are (so it seems) appeals merely to compassion; and whenever human welfare, even of a trivial sort, might be promoted by doing harm to animals, there is no serious constraint on such conduct (it seems) if animals do not have rights.

II

In this and the next two sections, I discuss a view of justice and injustice, that of Simone Weil, which is deeply opposed to contemporary thought about how justice is linked to rights.[2] I shall then turn back to discussion of animals and injustice.

The difference between justice as Weil understands it and justice as we usually understand it in terms of rights is—putting it in Wittgensteinian terms—a difference in grammar. In discussing the contrast between her view and the usual view, I have a central idea of Wittgenstein's in mind: that differences in grammar reflect differences in what we find important.[3] Weil's writings on justice and power articulate a responsiveness to the relentless treatment of vulnerable human beings. She invites us to share (or to recognize that we do share) her sense of the significance of evil done to the vulnerable, and she tries to show that the grammar of justice, if it is tied to rights, obscures the difference between such evil and other sorts of treatment to which human beings may be subjected. A grammar that obscured or overlooked that difference would suit people for whom nothing hung on the difference, and Weil suggests that the Romans were such people. Her description of a contrasting grammar of justice, not tied to rights, shows us a different way of making sense. I spend as much time as I do on Weil because I think that what underlies the animal rights movement is a responsiveness to the vulnerability of animals in the face of the relentless exercise of human power, and that the articulating of that responsiveness calls for a grammar akin to the grammar of justice as Weil describes it.

Weil takes the concept of justice not only to be distinct from that of rights, but to come from a different conceptual realm. She argues that, when genuine issues of justice and injustice are framed in terms of rights, they are thereby distorted and trivialized. Our grasp of the evil of injustice is impeded when it is spoken of as violation of rights. The language of rights comes down to us from the Romans; rights in the original sense were, she notes, rights to property—and property centrally in slaves.[4] The language of rights was shaped in that conceptual world and still bears its stamp. It is eminently suitable for complaints that I am getting less, for example, than I am entitled to for something I want to sell, but not for the expression of outraged hurt when real evil is done to a person. The cry of hurt of someone to whom evil is done is altogether different from someone's outrage at getting less than what he takes to be his fair share of something.[5] Weil does not deny that there are contexts in which not giving someone his or her fair share may be a case of real injustice to that person. But what would make it a case of injustice in her sense cannot be explained in terms of the unfairness of the share.[6] The capacity to respond to injustice as injustice depends, not on the capacity to work out what is fair, but on the capacity really to see, really to take in, what it is for a human being to be harmed. This is not easy for us; it requires a recognition of our own vulnerability, and there are no comparable demands on us in thinking about deprivation of rights.

Weil claims, then, that the attempt to give voice to real injustice in the language of rights falters because of the underlying tie between rights and a system of entitlement that is concerned, not with evil done to a person, but with how much he or she gets compared to other participants in the system. She criticized, for example, the use of the language of rights in formulating demands on behalf of workers. The dignity of physical labor was among her particular concerns, along with the fact that in modern conditions—in factories speeded up to maximize productivity—physical labor is degraded. And she takes that to be a serious injustice. But, when trade unions speak for workers, what they demand for the workers is, not an end to such conditions, but *higher wages*, a right to a larger share. Framing demands in that language makes it impossible to see what should be the object of their concern: how work in modern conditions damages those who must do it. She says, "Suppose the devil were bargaining for the soul of some poor wretch and someone, moved by pity, should step in and say to the devil 'It is a shame for you to bid so low; the commodity is worth at least twice as much'." [7] The language of rights fits the context of economic bargaining and economic demands; within such a context I may claim that I have a right to more than you are giving me. But Weil wants to contrast cases of economic demands (and, more generally, claims of fair entitlement) with (for example) the desperate cry that might be made by a girl being forced into a brothel. This fierce desperate cry at what is being done to her, the response to the harm done and further harm threatened, comes from the depth of her soul. If the word "rights" is used to describe her situation, the case is, Weil says, falsely assimilated to that in which a farmer, browbeaten to sell his eggs at a moderate price, responds, "I have the right to keep my eggs if I don't get a good enough price." [8] The language of rights is the language of a "middle" or mediocre level of values; such language cannot express genuine injustice or the needs of those who are its victims. The distinction between two "levels" of value, and the connection of rights with the lower, "mediocre" level and justice with the higher, are at the heart of Weil's treatment of justice.[9] The distinction between appeals to justice and appeals to rights is for her also tied to a difference in the response elicited by the two kinds of appeal; see note 3 below.

Weil makes another important point: that in a context in which we *do* assert our rights, there must be force in the background, force that we can rely on, or else such a demand will be laughed at. Thus it may be perfectly true that I have a right to the coat that you are pulling off my back. But if I demand my rights in a situation in which *you* have all the power, you can laugh at my demand. And, indeed, if someone is threatening to take your life, what is the idea that, if you have a right to something, you can *demand* it, rather than merely beg for it? As the Athenians in the Melian Dialogue remind the Melians, the Melians are in no position to demand their rights. They do *ask* the Athenians to make a treaty with them accepting Melian neutrality, and they suggest that, if the Athenians attempt to subjugate them by their superior force, this will in the end have bad consequences from the point of view of the Athenians' own aims.[10] The Melian strategy here is far from unusual. When threatened by injustice, we may attempt to show the person that, if he acts unjustly, this will in the end lead to his being harmed in some way. And, as in the case of the Melians, we may well be bluffing.[11]

We should also note that, when we are concerned about injustices to third parties (for example, if we were to write to the Minister of Justice of a country unjustly imprisoning someone) we do not say "I demand that he be given his rights." Or we can say it, but we should know that our demand will be ignored or laughed at unless we are pointing some large guns at his capital city, or are ready to end his most-favored-nation status in trade, or whatever. This does not mean that there is no language we can use, or that we have only the possibility of begging for kindness. Kindness is not at issue, justice is; but that does not mean (on the Simone Weil argument) that in such cases the language of rights is what is needed.

There are important insights in Weil's view. I am concerned here with three, namely those into (1) the existence and importance of the distinction between the moral level of concern with rights and the moral level exemplified by Weil's cases of genuine injustice; (2) the falsity of the idea that we can object to injustice only in the language of rights, supposedly because rights can be demanded (so Weil's insight is into the falsity of the idea that the alternative to demanding rights is begging for mere charity, mercy, or kindness, and thus letting justice slide out of the picture); and (3) the fact that tying justice and injustice closely to rights encourages misunderstandings of the complex relation between justice and compassion, and between acting unjustly and being pitiless. What does it mean to call these *insights*, if I say that the difference between justice as Weil conceives it and justice thought of in terms of rights is a difference in grammar? A full discussion would go too far afield; here I want only to note that what appears from one point of view to be moral insight appears from another to be a kind of *grammatical description or redescription*.[12]

III

This section is about some objections to Weil's view.

It may seem unreasonable to suggest, as she does, that the language of rights has *remained* tied to the conceptual world from which it emerged, in which property rights included the right to own people. Before confronting this issue, one should, though, note her discussion of the use of the language of rights by the "men of 1789."[13] Although the world of 1789 is no longer the ancient world, it is still a world within which slavery is widespread, and within which there can be no end to slavery without either the violation of property rights or the indemnification of slave owners. The rights of the owners of human beings, their property rights, shape the kind of problem slavery is within that world. Or, again, think of the 1840s, and imagine that someone then believed that people in Ireland at the point of death by starvation and disease had some right to life, and that therefore food supplies held by the various relief committees should be made available to them, i.e., that the committees had, derivatively, a right to distribute supplies of food to the starving. Such action was, however, ruled out by the authorities because making food available to the starving either free or at a price they could afford would lower the general level of prices and thus reduce profits; it would thus interfere with property rights. (The Dublin Relief Commission, for example, was, in December 1846, receiving more than twenty-five letters daily from relief committees, asking for permission to sell food cheaply. These requests were invariably denied, although prices were in fact going sky high and speculators were making fortunes. The official view was that "the famine . . . offered traders an opportunity to make profits, of which it would be unjust to deprive them.")[14] The point here is not that property rights invariably trump other rights, but that rights remain within the sphere in which we are sharing things out, exchanging this for that, balancing this right against that.[15] We can note that property rights are still frequently treated as framing the constraints for dealing with problems involving grave injustice, and in that way as fundamental, for example in social and economic policies dictated by structures of indebtedness between developed and underdeveloped countries.[16]

The attempt by Weil to distinguish injustice from violation of rights, and to characterize rights as remaining within a sphere of contention, of claims and counterclaims,[17] might be said to go wrong because it implies that genuine injustice is somehow beyond a sphere of opposing claims; yet, surely, there may be occasions on which avoiding injustice to some can only be at the cost of injustice to others. There can be no sphere, it may be said, entirely above such conflicts. What Weil says does allow for there to be conflicts in which there is justice on both sides, while misunderstandings obscure the character of the conflict.[18] In what follows, I assume (though I think it doubtful) that she does imply that conflicts that do not involve misunderstandings, and in which neither side can be treated justly without injustice to the other, cannot arise. How, then, does this matter for what we can learn from her?

On behalf of Weil, one might in any event note that the framing of social questions in terms of rights makes it appear more frequently than it is the case that in some particular situation we must unjustly harm some if we do not unjustly harm others. The *character* of our conflicts is made obscure when two sides of a conflict involving very different elements of human life are expressed in the same terms, as in the case in which Irish victims of a profoundly unjust social system were to be allowed to starve because distributing food cheaply would interfere with the rights of traders to make a high profit out of the famine, and would thus supposedly be an injustice to them. Here the understanding of the traders as possible victims of injustice depends on starting from a conception of their property rights, taken to include rights to speculative profits. The framing of thought about the conflict in terms of the rights of the parties is thus likely to lead to a misconception of its character. This possible obscuring of what is involved in a situation is illustrated also by Weil's imagined example of someone attempting to protect a person for whose soul the devil is bargaining. Thinking of that situation in terms of rights to something property-like leads naturally to a misconceived solution: make the offer higher. Pay higher wages to those whose lives are deformed by the conditions in which they have to work; make the bargain better; treat their property in their labor as worth a little more. The situation is understood in terms of contending economic interests; and injustice in Weil's sense disappears from view. That is to say that the evil done to those who are the victims is what we do not really see.

But even if we grant to Weil that conceiving a situation in terms of rights may obscure its character, we have not adequately answered the original objection: that talk in terms of rights does not have to retain the conceptual connections exemplified by Roman rights to property in human beings. An expression of that objection might go this way:

> Whether or not the concept of a right comes to us from the sort of Roman context described by Weil, we should be concerned with its present use and its present significance in social and political discussion. In our present political arguments (and in political arguments of the last two centuries) it is often used so that there is a contrast between institutionally established rights, like rights to property in human beings, and human rights. Compare the notion of a *gift*. It is conceptually open in something like the way the notion of a *right* is: one can make a gift to someone of a slave, for example, and yet we can find examples of talk of gifts in which the notion of a gift attaches to human goods of the greatest significance and is used to bring out the significance of those goods. The description of human liberty as a gift from God can thus play a role in attacks on slavery as unjust deprivation of liberty. And just as the notion of liberty as a gift from God can have a role in defending human beings from injustice, without distorting or trivializing that issue, so can the notion of a right.

That argument is meant to take us to the conclusion that talk in terms of rights is not by its very character likely to distort or trivialize a claim involving genuine injustice.

The argument is important, not in that it is conclusive, but in that it moves us further into the questions. Here we should consider what might at first appear a good counterexample to Simone Weil. In the middle of the eighteenth century, the Quaker John Woolman wrote two very moving attacks on the evils of the slave trade, the practice of keeping slaves, and the racism underlying the trade and the practice.[19] He does refer to a natural right of human creatures to liberty,[20] and the case seems to be an excellent counterexample to Weil because his use of the language of rights does not trivialize the issues. But we should look at the essays themselves to see why it does not. Woolman's arguments in no way depend on the notion of rights, which in fact plays very little role in either essay. The arguments of both essays are designed to bring to the consciousness of Woolman's fellow Quakers and other Christians the injustice of participating in the slave trade and of holding slaves. The power of the arguments, especially in the later essay, comes from imaginative descriptions making evident the injustice of the capture and transportation of slaves. Woolman attempts, in several very moving passages, to make his readers understand the "inexpressible Anguish of soul" of those who survive, uncaptured, an attack by slavers, and also of the captives themselves; he brings out the self-deception involved in distinguishing between actually going out and catching and stealing people and merely owning such stolen people or their children; he uses Biblical texts superbly, especially passages from the Hebrew prophets, and he makes it hard to see those passages in any other way than as condemning acts and practices resembling those of modern slaveholders. *These* arguments bear the weight of the work; they open the reader's eyes to the cruelty and injustice of the slave trade and of continuing to hold slaves and live on their labor. The meaning of a reference to "the right to liberty" in *this* context is given by the surrounding text. The injustice involved is not explicable as the violation of a right; rather, the moral force of Woolman's brief references to a right to liberty is inseparable from his understanding of the kind of life for which human beings are intended, and of the afflictions and anguish visited on the victims of slavery.[21]

Woolman's argument is in a sense the reverse of the kind of argument that led to the conclusion that it would be unjust to traders to distribute food cheaply to starving people in Ireland. There, the notion of injustice to the traders depended on an understanding of their nonlegal rights to make high speculative profits, whereas in Woolman's writings, the vivid portrayal of unjust treatment stands on its own, and the understanding of injustice gives Woolman's references to rights the kind of moral seriousness they have. In Weil's terms, what Woolman does is help his readers to hear the cry "Why am I being hurt?" The starting point is crucial: the opening of sensibility to the affliction of the captured Africans (and of their families, and of slaves born in the Americas). The argument's understanding of injustice comes from that sensibility, which is meant to lead its readers to question how they can continue to be parties to the imposing of such affliction. What Weil criticizes is the mode of thought in which the direction of understanding is *reversed*—in which the notion of rights shapes our understanding of injustice, conceived thus in terms of what is owed to a person compared with what is owed to others. So, what the example of Woolman's essays shows is that, if we are to interpret Weil reasonably, we will not take her to hold that the mere occurrence of the language of rights makes it impossible for social argument to do justice to injustice.

(We should note here that much of the history of this country is characterized by the use of arguments moving in the direction opposite to Woolman's, i.e., starting from rights. This is particularly evident in arguments that have gone from rights to liberty to the right not to be interfered with in practices involving serious injustice—injustice to slaves, for example, and later to their descendants. What goes wrong in all such accounts, on Weil's view, is ultimately a matter of the connection between the notion of rights and a conception of autonomy linked to what she describes in "Human Personality" as the "personal.")

IV

In this section I am concerned to avoid some possible misunderstandings of what Weil says about rights.

Weil is not denying that one may have a legal right that this or that form of unjust treatment not be done to one. This is obvious from one of her examples, that of putting out someone's eye, which would be a case of injustice on her view and would also violate rights to bodily integrity protected by criminal law. Nor does Weil want or need to deny that the existence of such legal rights may be of great importance in preventing injustice. Rights can work for justice or for injustice; the concept of a right has, on her view, a kind of moral noncommitment to the good. It is thus a concept unlike justice. This difference is then part of the overall contrast between the "level" of justice and the "level" of rights. Justice and respect for rights, as concepts, have different sorts of relation to good.[22]

So Weil is not denying that one may have a legal right that some form of injustice not be done to one, nor that such a right may be significant in preventing injustice. Nor is she denying that among the legal rights that may work to prevent serious injustice are property rights. An excellent example would be the existence of forms of tenant rights. While oppression of peasants in many places has been furthered by laws allowing landlords various rights, in many places it was to a degree impeded by the existence of certain rights of tenants, including rights to glean or to graze animals, or rights providing some security of tenure or preventing the annexation by landlords of improvements made to the land.[23] So some property rights, like other rights, can on the Weil view be parts of systems of unjust oppression, while others may provide some security against unjust oppression. The "right to property" is not one right: property rights considered merely as such share what I called the moral noncommitment to the good that belongs to rights in general. (See also Weil's discussions of the need for personal property, not including money but including things like a house or a field, furniture, and tools.)[24]

While it is true that Weil treats issues of justice as having a depth not present in issues of rights, she is not denying the importance of dealing adequately with problems of rights. She says somewhat sneeringly that minds capable of dealing with these latter problems can be formed in a law school.[25] We should note here how the case of slavery illustrates what she means but also shows a weakness in her account. Slave narratives and writings like Woolman's enabled many white people to see and to take seriously the injustice of slavery in this country. But there were many apparently insoluble practical questions about how those who held property in slaves were to be treated, if the institution of slavery were to be ended without revolution. It is all very well to sneer at the sort of mind capable of dealing with questions of fair legal compensation, but the apparent insolubility of such questions can help to maintain injustices for years. It is no essential part of Weil's view, though, to run down the kind of difficulty such problems present.

We can recognize and retain important insights in Weil's view even if we drop her idea that contention is characteristic of the level at which rights are claimed, and not (except when there is some misunderstanding) of the level at which we are concerned with justice and injustice. If we accept her distinction between rights and justice, between the language of objection to interference with rights and the language of outrage at injustice, there will be a significant consequence. It will follow that there is something wrong with the contrast, taken to be exhaustive, between demanding one's rights and begging for kindness—begging for what is *merely* kindness. The idea that *those* are the only possibilities is (as I noted) one of the main props of the idea that doing injustice *is* failing to respect rights. It seemed as though, if you were not claiming that you had a right not to be treated as you were being treated, or if you could not be interpreted as making such a claim, you could not be crying out against a real wrong done to you. On the Weil view, the response to injustice being done to one is *neither* an appeal for one's rights *nor* a complaint that one has not been treated charitably.[26]

V

I have not tried to give a full account of Weil's views on injustice, but simply to present a conception of injustice at some distance from much contemporary thought. I turn now to the question how her ideas might bear on our relation to animals.

At first sight, there appear to be such great obstacles to connecting Weil's thought with animals that it might seem merely perverse to try. Animals hardly appear in her writings, which bear a clear impress of a Cartesian culture. When she asks if there are obligations *to* anything but human beings (within the sphere of nonreligious obligations), the only candidate for object of obligation other than human beings that she considers is *collectivities* of human beings. Animals do not come in even as candidates.[27] When she speaks of base and cruel acts, her examples are base and cruel acts to human beings; it is not clear what sort of room she would allow for actions being base and cruel to animals. (I don't mean that she would not allow such a possibility, but that it illustrates her mode of thought that the question is not one that she sees as arising.) She has a Cartesian sort of interest in questions like whether language provides *the* difference between animals and us, and is inclined to observations about animals that seem to have no empirical basis. She remarks, for example, that nothing individual and concrete exists for animals.[28]

But all that is less important than her approach to justice itself, which seems to rule out any application of her views to animals other than to suggest that there can be no such thing as injustice to an animal. She writes that "at the bottom of the heart of every human being, from earliest infancy until the tomb, there is something that goes on indomitably expecting, in the teeth of all experience of crimes committed, suffered, and witnessed, that good and not evil will be done to him."[29] This profound and childlike expectation of good in the heart is not involved when one demands one's rights, but is the source of the cry of outraged hurt when injustice is done to one. And, if the idea of a human being as a possible victim of injustice is inseparable from the expectation in the heart that good will be done to one, how can animals be thought of as possible victims of injustice, since animals do not have this expectation of good?[30] It thus looks as if, if Weil's account of justice and injustice has any bearing on our relation to animals, it will simply rule animals out as possible objects of unjust treatment. It may also seem that the concept of *rights* is better suited for attacking bad treatment of animals than is that of injustice, if injustice is conceived along Weil-like lines.

But things are not as simple as they seem, and the implications of Weil's account cannot be read off it so easily.[31] We should note first something very striking in the way she relates justice to our expectations. It is quite different from what one might take to be a natural move here. One might, that is, think that to treat people unjustly is to treat them worse than they have a right to expect, the basic notion in just treatment being then what one may reasonably or legitimately expect, or what one has a right to expect.[32] A great difference between any such account and Weil's is that the basis of justice on Weil's account is not that it is reasonable for us to expect good to be done to us, or that we have a right to expect good to be done to us. The expectation is in a sense not reasonable at all. And the part of the heart within which there is this expectation may be almost numbed by repeated blows and the infliction of much suffering; yet the expectation is still there even if the person is no longer capable of crying out. This *un*reasoned, strangely persisting expectation is what can, if we are willing to attend to it, stop us from inflicting harm. Such inhibition, as a response to the expectation of good, is not respect for what the other person rightfully expects, and is not itself a kind of response for which there is available a reasoned justification. The basis for justice, on Weil's view, is not just the existence in us of an unreasoned expectation of good, but, equally, the possibility in us of being brought up short by that expectation, of being touched by it, of finding ourselves reluctant to go ahead with harm that can elicit from that place in the heart of the other person the cry "Why am I being hurt?" Weil is not suggesting that it generally or even frequently happens that people *are* thus inhibited. Treating people as mere puppets can be exhilarating; it can be enjoyable to make people cry out with outraged hurt, and very often we are so unaware of the response of the victims of injustice that we are easily able to go ahead with what we are inflicting on them despite their internal lamentation. Weil's writings about responsiveness to injustice connect directly with Woolman's writings on slavery. He is trying to get his readers to attend to the anguished hearts of those taken as slaves and of the members of their families. What Woolman hopes is that the possibility of our being brought up short by hearing and attending to the anguished cries will be realized in his readers.

Weil's understanding of justice thus depends on two things: first, that there is an unreasoned expectation that good and not harm will be done to one, and second, that there is an equally unreasoned possible response to that expectation, viz., an unwillingness to go ahead and do harm in the face of the expectation, a response that may also include a desire to protect the being who has the expectation.[33] The awareness of the other being that impedes doing injustice, doing harm, is a kind of love, or loving attention; and this then brings out in another way how far Weil's approach to justice is from that in contemporary moral theory. Contemporary thought about justice pushes apart justice, on the one hand, and compassion, love, pity, tenderness, on the other; but Weil's conception of justice has at its center the idea that a kind of loving attention to another being, a possible victim of injustice, is essential to any understanding of the evil of injustice. This point of hers fits closely with the idea that the level of justice is distinct from that of rights; justice conceived in terms of rights can be divorced from and contrasted with love and compassion. This separation is the source of enormous amounts of discussion in contemporary moral philosophy about an ethics of justice versus an ethics of care. The Weil response to that whole discussion would be that the very idea of such a contrast and distance between justice and care or love is one of the bad effects of thinking about justice in terms of rights.[34]

VI

It may seem that we are no further forward in our attempt to see whether Weil's thought about injustice might have a bearing on our treatment of animals. That is, I first explained one idea that seemed inapplicable to animals, that of a profound and unchanging expectation of good in our hearts, and now I have added to that a second idea that seems equally inapplicable to animals, namely, that of a response in us, an inhibition that can be called forth by attention to the other person and awareness of that expectation of good.

But we have got further than appears. We have in Weil's thought an understanding of justice that lacks the structure of theories of justice in contemporary philosophy; and it is this lack that will be important for us. We can see the kind of structure that is absent by noting how rights theories work. They have *this* structure: a being *can* have rights because it has certain morally significant characteristics. Because it has these characteristics, it has rights that we as moral agents ought to respect. The morally significant characteristics that are treated as the basis of rights, the characteristics that make a being a possible bearer of rights, vary according to the theory. In some theories, the basic characteristic is having interests (interests themselves, on some accounts, being held to be dependent on desires); in others, it is having "inherent value"; in others, it is being oneself a moral agent; and so on. For any such theory, the question will then arise whether animals have, or have literally, that characteristic which, according to the theory, makes it possible for a being to have rights. And thus theories with this sort of structure invite the kind of debate that has become very familiar, as attempts are made to prove that animals do have such-and-such, or that they do not, or that, although they do not, the such-and-such is not what is requisite for having rights but some other such-and-such is.[35]

It may seem as if Weil's account has a similar structure, since she connects our being possible victims of injustice to our having an expectation of good. So it looks as if, according to her theory, the crucial characteristic that makes a being worthy of moral respect is having that expectation. (And the idea that her account has that structure may be reinforced by her presentation of how the expectation can elicit a response from us. It may seem that the response is a recognition that the being whom we might harm has a characteristic in virtue of which we ought not to treat him or her that way; and so it looks as if Weil is committed to an account holding that all beings with x are entitled to respect.)

To see Weil as having such an account is to read her work with preconceptions shaped by our idea of what a moral theory needs to do: it needs to show what sort of beings it is rational to treat with respect. But that is not what she is doing. If she were giving us a moral theory of the same general type we are familiar with, it would indeed imply that justice and injustice have no application to animals. And the appearance that it has that consequence reflects a natural enough inclination to read her as putting forward such a theory. It would, though, be a peculiar theory of that sort: why pick on the expectation that good will be done to one as that characteristic on which being a possible victim of injustice depends?

If we note what a peculiar theory it would be if it were a theory, we can perhaps grasp more clearly how her views do hang together. What can be awakened in us by awareness of a human being's hope that good will be done to him is the desire not to dash that hope, not to violate what it springs from; both the hope and the desire responding to it are elements in human attachment to the good, and are on the same level. Another important part of the Weil conception needs also to be mentioned if we are to see how her views hang together. She is deeply impressed by the relentlessness of those who do exercise power over human beings, and impressed by how easy it is, indeed, to treat human beings as mere puppets, mere things. What she means, when she writes about loving attention to human beings, includes a response to human vulnerability in the face of that relentlessness.

What, then, is involved in thinking of injustice, in anything like Weil's sense, as applicable to animals? I suggest that we think of this as a response to communicative pressure; so I need to explain what I mean by that. In our various activities, including our attempts to think about our lives and to make sense of what we experience and what we do, we can use words well or badly. The ways of speaking we find in response to activities and experiences may accommodate a merely superficial kind of "meaning it"; or we may be able to find words that more fully render experiences or activities, words that can be meant more fully. (And it also works the other way: we can say something, and only later find what activities or experiences are involved in meaning fully what we had said, as Horton the elephant finds out over a long time what is involved in having meant what he said.)[36] Who then is to say what ways of speaking respond well to communicative or expressive pressure? Philosophers often rush in to judge such ways of speaking on the basis of rules and regulations. You cannot say such-and-such and mean anything, we say, because what you say is not verifiable, or because what you say is not part of the language-game in which the words you use belong. And those of a scientific temperament may also rush into judgments of this sort.

A good example of what I mean by communicative pressure is the use by animal trainers of apparently anthropomorphic language in connection with their work as trainers. Vicki Hearne uses such language herself, but also discusses its role in connection with the activities and lives and experiences of trainers, and the pressures on them from outside to reject such ways of speaking. But the question is how far they can mean what they say, in lives formed in part by such ways of speaking; and no outside attempt to lay down rules for what can be said about animals can take into account the pressures from within that life to speak as they do, to *make* sense of what they say within their lives. Hearne argues that what it is to mean what the trainers say cannot be seen in abstraction from the work of language in making possible the occurrence of the animals' work.[37] In the background here is the fact that Hearne is by trade a poet as well as an animal trainer, and being by trade a poet is being concerned by trade with what it is to mean one's words further and more fully.

This argument applies also to Weil's own account of justice and injustice. She believes that, in the human expectation that good, not evil, will be done to one, we are in contact with the Good. And this idea may seem to require a metaphysical theory about the Good with which we are supposedly in contact; it may seem also to make Weil vulnerable to the charge that we cannot (in the face of Darwin and contemporary science, or of other intellectual and cultural developments) really believe in such a Good anymore. But such criticism ignores the communicative/expressive pressure to which Weil is responding. Here I shall quote from "One More Day" by Czeslaw Milosz, a poem which is about the communicative/expressive pressure to speak of Good, and which is also itself a response to such pressure.

> Comprehension of good and evil is given in the running of the blood.
> In a child's nestling close to its mother, she is security and warmth,
> In night fears when we are small, in dread of the beast's fangs and in
>   the terror of dark rooms,
> In youthful infatuations where childhood delight finds completion.
>
> And should we discredit the idea for its modest origins?
> Or should we say plainly that good is on the side of the living
> And evil on the side of a doom that lurks to devour us?
> Yes, good is an ally of being and the mirror of evil is nothing,
> Good is brightness, evil darkness, good high, evil low,
> According to the nature of our bodies, of our language.[38]

Talk of good and evil has its modest origins in closeness to one's mother and in childhood terrors—but it does not need a more metaphysical basis. Attachment to the good, and the idea of good as linked to being, and evil to nothingness and darkness: these ideas need no metaphysical justification. Although many people do take for granted some or other way of thinking that does not link Good to Being, it is a mistake to think that we now have some basis for ruling out the kind of talk about good and evil that is central to Weil's thought about injustice.[39]

I am arguing, then, that Weil's thought about injustice should be understood as a response to communicative pressure, pressure within her life and experience, a response especially to impressions made on her by human vulnerability and hope and by the exercise of power without pity. Then the further suggestion I am making is that we can see a comparable kind of communicative pressure toward connecting Weil's conception of injustice with much of the treatment of animals. That is, it is pressure to extend something like her conception of injustice to animals; hence it depends on the sense of injustice, the sense of good and evil, that Weil writes about in connection with human beings. It involves a comparable horror at human relentlessness and pitilessness in the exercise of power; it involves also horror at the conceptualizing of animals as putting nothing in the way of their use as mere stuff. Just as Weil's language responds to her sense of the life of human beings, and of the connection between that life and the Good, so the communicative pressure to extend talk of injustice to animals responds to a sense of their life, and a seeing of a connection between their lives and the Good. In both cases, the idea is that attention to these lives, seeing their connection with the Good, is capable of stopping us from treating them as props in our show.

(I should perhaps specify here something I am not saying, and indeed am rejecting. There is an idea that human beings have "intrinsic goodness" or some such thing, and that this can serve in a theory as a basis for according us rights, treating us with respect, or whatever, and then there is the question whether animals have a similar metaphysical status, and, if so, whether it can serve in a theory as a basis for according them rights or whatnot. (This is a notion of "intrinsic good" that is intended to be thicker than, and to do more work than, any notion that merely reflects the fact that we may regard it as good to do things for a human being or an animal, things that redound to its benefit, without trying to achieve something else, something external to the good of the human being or animal.) I am not making any metaphysical claims about people or animals. It seems to me that the metaphysical claims issue from one way of seeing the phenomena here, the phenomena of our making connections in thought and talk between human life and the Good, or between human and animal life and the Good. I am trying to provide an alternative way of looking at those phenomena.)

VII

In this brief section I discuss an interesting (partial) parallel between the cases Weil has in mind and a kind of case involving animals. In "Human Personality," Weil twice puts before her readers the image of a poor inarticulate man, accused of stealing a carrot, brought up in front of a magistrate who keeps up an elegant flow of queries and witticisms, while the poor man is unable to stammer a word.[40] This illustrates exactly what I meant by our treating someone as a prop in our show. But animals also are treated in this way. Some videotapes stolen from the University of Pennsylvania Head Injury Laboratory contained a section which many people found particularly shocking, in which some of the staff were seen ridiculing one of the baboons.[41] This sort of subjection of an animal to one's sense of humor, making it into a prop in one's jokes at its expense, is a particular way of exercising power over it. The animal's body, which is all it has, as a poor man's body may be all he has, is turned into the mere butt of your jokes. The butt of your jokes cannot resist you; this is part of the fun of it. It amuses us to turn a living animal, which lacks the power to get away, or to resist, into an opportunity for jokes. The moral disgust with the antics in the Head Injury Laboratory may be compared with the response of the Dayaks in Borneo to ridiculing or humiliating an animal, to dressing it (for example) in human clothes "in parody of humanity." One must not, they believe, make fun even of a fish or a frog; this is a great crime, on the level of incest.[42]

The discussion of animal rights almost invariably leaves out the ridiculing of animals. Those who ascribe rights to animals take those rights to be dependent on interests, and the interests of animals are thought of as dependent on what the animal might be aware of, or on what might in a naturalistic sense be taken to be part of the good of the animal. Not being a butt of humor is not taken to be part of its good, and so the whole issue of ridicule of animals is left out of the discussion of rights, despite the fact that many people respond very deeply to such ridicule as among the appalling things that we do to animals. Animal rights theorists are also very suspicious of cases in which it might be arguable that we ought not to do something to animals but it might also be said that the duty in question is not a duty *to the animals*. It may seem to a defender of animal rights that, although we should not ridicule animals, it will be hard to make out that we owe it *to the animals* not to ridicule them. In the background here is the idea that we cannot owe it to animals that they not be treated in some way, unless they would *suffer* from such treatment; and the idea would be that an animal cannot suffer from being ridiculed if it is not even aware that it is being ridiculed. Ridicule is not seen as something we are really doing *to the animal*. This is, however, a philosophical interpretation that we impose; the notion of what it is for a person or an animal to be the victim of our ridicule can have various shapes; and we do not have to shape the concept of "victim of ridicule" so that the animals (or people unable to comprehend that they are being made butts of ridicule) do not really count as *victims*. In the case of animals, and of people unable to realize that they are being made the butt of ridicule, the powerlessness of the victim is part of what is enjoyed by the ridiculers, and the powerlessness includes the victim's incomprehension.[43]

VIII

In this section I look at some of the differences between thinking of animals as having rights, and thinking of them as possible victims of injustice in something like the sense explained by Simone Weil. I start from the idea of being impeded from some action, which has a place both in theories of rights and in Weil's views. There is similarity here, but also an important contrast.

The point of ascription of rights, we are sometimes told, is precisely to block some paths by which we might otherwise seek to maximize welfare or in other ways to achieve our ends. The idea is that respecting rights essentially can involve making us stop doing, or refrain from doing, what it might otherwise seem desirable to do. The police cannot simply walk into your house and search it, even if it might appear very desirable to do so from the point of view of the general welfare, because you have rights against such interference, protected by law. Weil's writings on injustice also give a role to the impeding of actions that one has the power to carry out, but the character of the impediment is different, and is directly dependent on a sense of good and evil, and on a response to the reality of the other human being. It is that person's connection with the good that works as the impediment.

Earlier I noted that it is not impossible for talk about rights to be informed by attention of the sort Weil discusses. That was the point of the Woolman example; Woolman refers to rights, but in a context in which he is trying to enable his readers to attend to what slavery is for its victims.

The protection afforded by rights does not, however, depend on achieving the kind of attention Weil speaks about; and talk of rights is frequently motivated in part by a desire to secure certain interests without appeal to anything like Weil's conception of Good or of loving attention (such attention being something we can by no means count on). Without such attention, though, there is no perceiving the evil of genuine injustice; and that is why describing cases of real injustice in the language of rights can distort those cases. The language of rights is, one might say, meant to be useful in contexts in which we cannot count on the kind of understanding of evil that depends on loving attention to the victim; this language is not meant to reflect differences that one can be aware of only through such attention. A language of another kind is what is needed, though, by the victims of such evil, or those who speak on their behalf.

In the case of communicative pressure to extend a Weil-like understanding of injustice to animals, what is central is a sense of the animal's life, the good of which is felt in the horror at what is imposed on the animal by our relentlessness.

Let me try to bring out more clearly what I am doing. There is not first an establishing that animals have characteristics that we share, and that are the basis for allowing animals to count as possible victims of injustice. Rather, there is a kind of response in the face of what is done to them: a pain and revulsion that requires for its expression the language of injustice, a pain and revulsion felt as akin to that at the exercise of power without curb over vulnerable human beings. This pain or revulsion or horror at what is done to animals has internal to it a way of understanding their lives, the reality of those lives, as tied to Good, and not just as tied to Good, but tied to Good in a way that seeks expression in the language of injustice. I have spoken of attention to the reality of those lives, but this use of "reality" is itself a response to the kind of communicative pressure about which I have been writing. After all, why should it not be "attention to reality" to note all the things that could be made out of the cut-up body of some animal? The *different* use of "reality" in question comes from the conceptual world described by Milosz, in which good is on the side of the living and is an ally of being.[44] To attend to the reality of animals, in this sense, involves seeing both the ways in which they are "with" us (an expression of which is "one breath permeates us all"),[45] and the ways in which they are strange and other. In discussing the Greek notion of what it is to "belong with" some other being, Richard Sorabji notes a point from Plato: that to "treasure others because they belong with us or are akin (*oikeioi*) . . . is different from treasuring them because they are *like* us"; we can "belong with" what is unlike us.[46] Such a combination of awareness of animals as "other" and recognition of their "belonging with" us is strikingly found in D. H. Lawrence: animals, who have their strange unknown lives, are inhabitants with us of this earth, linked to us by the "strange planetary phenomenon" of life. Lawrence remarks on the gift we are given when we are able to become aware of the "delicate realness" of these other beings.[47]

IX

Again in this final section I consider a treatment of injustice to people, and then turn back to animals.

In *What Then Must We Do?*, Tolstoy describes our responses to the unjust treatment of human beings, including those who carry out burdensome household tasks for us. He points out that we persuade ourselves that we feel for the suffering of those on whom the burdens of supporting us fall, and we try to make those burdens lighter—in any way except taking the burdens off them and bearing the burdens ourselves. We use a chamber pot, and we do not want to carry it out. So it is one of the things that our servants will have to do. Because we care for their welfare, or so we tell ourselves, we want to make this task lighter for them, and so we invent all kinds of devices to make it less of a misery—anything *except* the very simple device of carrying it out for ourselves. In a complex society, economic compulsion forces hundreds and thousands of poor people to labor in ways that make *our* life, the life of the well off, possible. We imagine we pity the poor, Tolstoy says, and that we wish to help them. He describes himself, and us, this way: I am on a man's back, choking him and making him carry me, and yet assure myself and others that I am very sorry for him and wish to ease his lot by all possible means—except by getting off his back.[48]

Tolstoy's point connects directly with current disputes about the treatment of animals—disputes between animal rights theorists and those who argue for some or other welfarist position—roughly, that we should avoid causing animals unnecessary suffering, avoid causing them suffering except so far as is necessary to the practices within which we use them. As is pointed out by proponents and opponents of welfarist views, what counts as unnecessary infliction of suffering will be determined by those engaging in the practices in which animals are used.

In terms of Tolstoy's images, we might say that the welfarist view is essentially that we should ease the burdens we impose on animals without getting off their backs, without ceasing to impose burdens on them, burdens that we impose because we *can*, because they are in general helpless.[49]

Welfarist views of how to treat animals may easily be combined with the idea that human beings are entitled to a kind of moral respect not owed to animals. The structure of such a view is then striking: we *can* make our respect-governed moral world possible, and enhance our own lives, through the burdens we place on the backs of animals, and we propose to do so, though the welfarist then adds that we should encourage the making of those burdens lighter by whatever devices do not interfere with the animals continuing to bear the weight we put onto their backs. The force of the animal rights movement comes from the sense of the profound injustice of this. It stresses two ideas: first, that animal welfarism leaves in place and unchallenged the unjust treatment of animals, and, second, that only an appeal to the rights of animals can serve as a proper alternative to welfarism. The demands of justice are then contrasted with appeals to kindness or compassion or care, because such emotional appeals are taken to lead to welfarism.

I have suggested that there is nothing the matter with conceiving justice as tied to pity. Pity may, at one level, motivate the welfarists whom Tolstoy describes; but Tolstoy himself does not contrast justice with pity. Instead, he leads us to see a kind of pitilessness at the heart of welfarism, a willingness to go ahead with what we do to the vulnerable, a willingness to go on subjecting them to our power because we can, because it suits us to do so, and it has suited people like us for millennia. "Willingness" is indeed too weak a word: we *will* not give up a form of life resting on the oppression of others; and the will to continue exercising power in such ways, the will to continue oppression, is inseparable from the "compassion" we express in welfarism. What Tolstoy shows is that there can be great dishonesty of heart in what we understand to be our own compassion. But this hardly implies that we need, or can find, some account of justice and injustice that severs the connection between justice and pity or compassion. Tolstoy brings to attention how deeply attached we are to the institutions that make possible all sorts of goods for ourselves at terrible costs to others, an attachment entirely compatible with attempts to make those costs a little less burdensome, and the institutions thus a little less troubling to more tender consciences.

In arguing against the theorists of animal rights, Bernard Williams criticizes the idea that there is "speciesism," analogous to racism or sexism, in the significance we give human beings in our moral thinking. Animal rights theorists make the mistake of trying to argue from a moral point of view which is not *ours*, which is not a *human* point of view; but, he says, moral arguments do have to be grounded in our point of view: moral reasoning does not drive us to "get beyond humanity." [50] The thinking about justice and injustice in relation to animals that I have presented is grounded in human moral thinking—in the perception of injustice in the subjection of others to our will. We mean to have a world in which we treat *each other* with respect, and we mean to make animals bear the burdens, the multiform burdens, of *our* living as we think human beings should. We mean to do this, and we have the power to do it. There is no need to see this from the point of view of the universe to perceive injustice in it.

Notes

I am very grateful to have had the opportunity to present an earlier version of this paper as the Julia Jean Nelson Rudd Lecture on Animal Rights at Indiana University. I profited greatly from the comments of the audience at that lecture.

1 G. E. Moore, "Wittgenstein's Lectures in 1930–33," in *Philosophical Papers* (London: Allen and Unwin, 1959), pp. 314–15, and Wittgenstein's discussion of "acceptance of a picture" at PI § 144. Cf. also the discussion of teaching someone the distinction between constructible and nonconstructible polygons by developing his understanding of "analogous" and "disanalogous," LFM, pp. 58–67. For a development of these ideas of Wittgenstein's in a somewhat different way, see John Wisdom on the methodology of "connecting" and "disconnecting," in "Gods," *Proceedings of the Aristotelian Society* 45 (1944–45), pp. 185–206.

2 I shall focus on the essay "Human Personality," in *Simone Weil: An Anthology*, ed. Siân Miles (New York: Weidenfeld and Nicolson, 1986), 50–78. Page references are to this edition, the pagination of which is different from that of the London edition of the same work. Weil discusses rights and justice in other works of roughly the same period; I shall not be concerned here with the problems of making clear how her views in the different works hang together.

3 See especially *Zettel*, §§ 378–88.

4 Ibid., pp. 61–62. On the ancient sources of the concept of human rights, see Richard Sorabji, *Animal Minds and Human Morals: The Origins of the Western Debate* (Ithaca: Cornell University Press, 1993), esp. chap. 11.

5 In the importance to Weil of the contrast between the two cries "Why am I getting less than he is?" and "Why am I being hurt?" we can see the influence of Rousseau's contrast, in his second Discourse, between *amour-propre* and *amour de soi*. Weil's conception of the cry "Why am I getting less than he?" as tied to what she calls "personality," and her conception of the second cry as tied to the "impersonal" follow Rousseau on the character of the difference between *amour-propre* and *amour de soi*. I should note here that the contrast between the two cries is not to be identified with that between the contrasting verbal expressions. There may well be occasions on which the cry "Why am I getting less than he is?" or "Why am I getting less than I have been getting?" may be expressed in the words "Why am I being hurt?" The important characterization of the cry against injustice is not in terms of the words used, but in terms of the heart's capacity to respond to real harm. That the issue is not a merely verbal one is shown by her remark (p. 73) that we need to learn to distinguish between the two cries. The implication is that this is not easy, which it would be if the difference were merely verbal. See also the discussion of the identification of the cry of injustice in Peter Winch, *Simone Weil: "The Just Balance"* (Cambridge: Cambridge University Press, 1989), pp. 181–83.

6 See Winch, *Simone Weil: "The Just Balance,"* chap. 14, esp. p. 181.

7 Weil, "Human Personality," p. 60.

8 Ibid., pp. 60, 63.

9 While there are numerous ways in which rights may be divided into categories, e.g., into basic and nonbasic rights, I do not think any such division corresponds to the kind of distinction that Weil makes. I cannot discuss this issue further here.

10 Thucydides, *History of the Peloponnesian War*, book 5, 84–113.

11 For her treatment of the Melian Dialogue, see Simone Weil, *Waiting on God* (London: Collins, 1959), pp. 98–101; also "Are We Struggling for Justice?," trans. Marina Barabas, *Philosophical Investigations* 10 (1987), 1–10.

12 See the sources listed in note 1 above.

13 Weil, "Human Personality," p. 60; cf. also p. 51.

14 See Cecil Woodham-Smith, *The Great Hunger: Ireland 1845–1849* (Harmondsworth: Penguin Books, 1991), pp. 167, 132.

15 Weil, "Human Personality," pp. 60–61.

16 It is interesting to note the contrast and similarity between Mary Ann Glendon's criti-

cism of the language of rights and that of Simone Weil. Both writers treat as central the distorting effect on political discourse of the model of property rights, but Weil sees this as permeating European rights discourse from the Enlightenment onward, while Glendon takes the property model to be peculiarly characteristic of Lockean-American thought about rights. The difference could be put so: for Glendon, the rights of 1789 could not be regarded as mere types of property right in the Lockean style (see esp. *Rights Talk: The Impoverishment of Political Discourse* [New York: Free Press, 1991], p. 35); for Weil, the language of 1789, although apparently about a diversity of rights, conceives them all in terms of the single underlying property model.

17 For the connection between claims of rights and contention, see Weil, "Human Personality," pp. 61, 63–64.

18 See Winch, *Simone Weil: "The Just Balance,"* on the role of misunderstandings in connection with Weil's views about justice. Winch, however, suggests that, on Weil's view, what justice in a particular case consists in can be arrived at only through discussion "in which all parties must be prepared to adjust their views" (p. 184). This seems to me questionable in connection with cases like that of the girl being forced into the brothel. Justice requires in such a case that this not be done to the girl. In order to see what justice requires here, we do not need to resolve any misunderstandings, nor would a discussion be needed in which both parties would have to be prepared to adjust their views. And, similarly, in Weil's case ("Human Personality," pp. 53, 71) of a magistrate who mocks an accused man. Winch attempts, I think, to give a single coherent account of Weil's views on justice as expressed in a number of different wartime essays; I would argue that Weil is pulled in different directions in the different essays, and gives a weight to consent in some of them that is not fully compatible with the treatment of justice in "Human Personality." One thing that would bring out the difference is that the account of justice in the other works cannot easily cover the unjust treatment of children, whereas there is no problem what would constitute unjust treatment of children on the account in "Human Personality." My reading of the different essays, put in terms of a Wittgensteinian conception of grammar, is that Weil develops several related but distinct ways of understanding the grammar of justice and injustice.

19 John Woolman, *Some Considerations on the Keeping of Negroes, 1754*, and *Considerations on Keeping Negroes, 1762* (New York: Grossman, 1976).

20 Woolman, *Considerations*, p. 39; cf. also p. 82 and *Some Considerations*, p. 17.

21 Woolman's argument also involves a connection between what Weil describes as the "impersonal" and genuine liberty; see *Considerations*, esp. pp. 82–83.

22 The issue here is complicated, and involves the idea of legitimate and illegitimate uses of a term. Thus, for example, consider again the claim that it would be unjust to the Irish traders not to allow them to realize speculative profits. Might this not show that justice has no more inherent commitment to the good than rights do? No, on Weil's view, because the notion of justice is not there legitimately used. Her idea, I think, is that the "middle" level at which the notion of rights operates means that it

can be used by the traders (or those who speak on their behalf) in claiming that they should be allowed to sell their stocks at high prices, whereas it is a misunderstanding of justice to attempt to derive it from considerations about rights. It is thus illegitimate to claim that distributing food to the starving in Ireland constitutes injustice to the traders by depriving them of speculative profits. See Weil, "Human Personality," pp. 76–77, on legitimate or proper use of words like "justice" and on the difficulties and dangers of using such words.

23 Rights protecting tenants were frequently customary rather than legal rights; hence the development of law, especially modern property law, frequently helped to clear the way for unjust treatment of those dependent on customary rights. Framing the issue entirely in terms of the rights on the two sides can obscure the issue of justice in Simone Weil's sense. Indeed, the poetry of dispossession focuses not on rights but on the pitiless treatment of the vulnerable, i.e., on injustice in her sense. To frame the issue in terms of rights on both sides would suggest a solution in terms of compensation, exactly the kind of solution that shows, Weil thinks, that what is at stake has been misconceived.

24 "A Draft for a Statement of Human Obligations," in Miles, *Simone Weil: An Anthology*, p. 209; also Simone Weil, *The Need for Roots: Prelude to a Declaration of Duties Toward Mankind* (New York: Harper and Row, 1971), pp. 34–35.

25 Weil, "Human Personality," p. 73.

26 A clarification: It may be asked what the relation is between the contrast that Weil emphasizes between justice and rights, and Aristotle's contrast between justice in a broad and justice in a narrow sense. Aristotle's "justice in the broad sense" includes all of virtue so far as it bears on one's relations to others. A man is unjust in this broad sense if he is an adulterer, whether through self-indulgence or greed; but if he commits adultery for money, and the motive is greed, graspingness, then he is unjust in the narrow sense, but the adulterer through self-indulgence is not unjust in the narrow sense. Injustice in the narrow sense is tied to unfair taking and to graspingness. The suggestion might be that injustice in this narrow sense corresponds to what Weil means by violation of rights, while injustice in the broad Aristotelian sense corresponds to Weil's "justice." But the suggested correspondence does not hold. Whatever is unjust in the narrow Aristotelian sense is also unjust in the wider sense; and this is enough to show that Aristotle and Weil are not making the same contrast. Further differences between the two contrasting pairs are worth noting. Many claims about what one has a (legal or nonlegal) right to do may reflect greed and graspingness; claims about one's rights to do this or that cannot be identified with claims about what one may in justice do, if justice is taken in Aristotle's narrow sense. (It is clear that, when the British authorities arrived at the view that the Irish traders had a right to realize huge profits, they would not have taken to be relevant to the issue of rights the fact that the traders were acting greedily and were seeking to get as much as they could, whereas those facts would be relevant to a claim about what was just, in the narrow Aristotelian sense. What is just, in this narrow sense, cannot be sepa-

rated from what a just man would do, and no just man would greedily seek the highest profits he could realize. This point does not depend on the rights in question being legal rights; they were not.) Actions from greed and graspingness will frequently be violations of justice in Weil's sense; they will also be violations of justice in Aristotle's narrow sense. So Aristotle's narrow sense of injustice does not correspond to the sphere of claims about rights, as Weil conceives the latter. And Aristotle's wide sense of injustice does not correspond to Weil's notion of injustice. If some action that is unjust in Aristotle's wide sense is unjust in Weil's sense, that would need to be shown in the individual case. Actions that are unjust in her sense would sometimes be unjust in Aristotle's narrow sense; they would also in general be unjust in Aristotle's wide sense, but Aristotle's category of injustice in the broad sense is wider. One reason, then, why the Aristotelian contrast and the Weil contrast do not coincide is that Aristotle's broad category is meant to be correlated to a broad category of virtues (and corresponding vices), Weil's to more specific virtues and vices (including the vices of pitilessness and of absence of attention).

27 Weil, *The Need for Roots*, pp. 3–9.

28 Because the falsity of this claim seems clear from even superficial observation of such familiar things as the attachment of dogs to their masters, there is a question whether Weil took observation to be incapable of falsifying such claims. See Simone Weil, *Lectures on Philosophy*, trans. Hugh Price (Cambridge: Cambridge University Press, 1978), p. 59.

29 Weil, "Human Personality," p. 51.

30 It can be argued that the generalization here is false, and that some animals do, for example, expect good from us. Thus, the fact that many dogs go on trusting and expecting good from human beings despite bad treatment shapes the moral revulsion felt by many people at what we do to them.

31 I have already mentioned that Weil's thought about justice and injustice was influenced by Rousseau's second Discourse, which is more obviously compatible with an application to animals than is Weil's own account.

32 See A. D. Woozley, "Injustice," in *Studies in Ethics*, ed. Nicholas Rescher (Oxford: Blackwell, 1973); cf. also Judith N. Shklar, *The Faces of Injustice* (New Haven: Yale University Press, 1990). Both Woozley and Shklar emphasize a distinction between expecting *that* someone will behave in a certain way and expecting *of* someone such-and-such behavior, although they do not formulate this distinction in exactly the same way. They both then explain injustice in terms of what we can expect *of* others. For Weil, in contrast, injustice is present when we go against people's expectation *that* good will be done to them, and this is not an expectation which can be shown to be legitimate.

33 See Weil, "A Draft for a Statement of Human Obligations," p. 203, for a description of the response as a matter of the heart's inclination.

34 See especially Weil, "Human Personality," p. 63: "If you say to someone who has ears to hear: 'What you are doing to me is not just,' you may touch and awaken at its

source the spirit of attention and love. But it is not the same with words like 'I have the right . . . ' or 'you have no right to . . . ' They evoke a latent war and awaken the spirit of contention. To place the notion of rights at the centre of social conflicts is to inhibit any possible impulse of charity on both sides." (It is important not to give the familiar narrow sense to "charity" in her remark.) See also Weil, "Are We Struggling for Justice?" p. 5: "One must be blind to oppose justice to charity; to believe that they have a different scope, that one is wider than the other, that there is a charity beyond justice or a justice falling short of charity. When the two notions are opposed, charity is no more than a whim, often of base origin, and justice is no more than social constraint. . . . Many controversies between right and left are no more than a conflict between a preference for individual whim and a preference for social constraint; or, more accurately perhaps, between dread of social constraint and dread of individual whim. Neither charity nor justice have anything to do with this." A contemporary thinker who has insisted on a connection between compassion and justice is Martha Nussbaum. Her account of the relation between compassion and justice is, though, quite different from Weil's. See, e.g., her "Compassion: The Basic Social Emotion," in *Social Philosophy and Policy* 13 (1996), 27–58. The idea of justice as linked to love in one's perception of others is also central in Iris Murdoch's thought. See Iris Murdoch, *The Sovereignty of Good* (London: Routledge and Kegan Paul, 1970).

35 For discussion of ancient theories of justice with a similar structure, see Sorabji, *Animal Minds and Human Morals*.

36 Some readers may not be familiar with Horton. See Dr. Seuss, *Horton Hatches the Egg* (New York: Random House, 1940). Horton commits himself to looking after a bird's egg; he meant what he said and he said what he meant (an elephant's faithful one hundred per cent); but only through many adventures and difficulties does he find out what is involved in meaning what he had said, and what the faithfulness is of which he spoke at the beginning, and of which he continues to speak.

37 See Vicki Hearne, *Adam's Task: Calling Animals by Name* (New York: Knopf, 1986), and "A Taxonomy of Knowing: Animals Captive, Free-Ranging, and at Liberty," in *Social Research* 62 (1995), 441–56. See especially p. 453, on "receptive Beholding," and p. 456, on belief prior to making what is said become truth and knowledge.

38 Czeslaw Milosz, "One More Day", in *The Collected Poems, 1931–1987* (New York: Ecco, 1988), p. 407.

39 I have in mind in this paragraph a variety of views that seem to rule out Weil's position. See, e.g., Christine Korsgaard on what "we" can no longer think. We can no longer, she says, link the real and the good in the way Plato tried to; rather, reality resists reason and value; and Kantian ethics is the only ethics consistent with the metaphysics of the modern world. See Korsgaard, *The Sources of Normativity* (Cambridge: Cambridge University Press, 1996), pp. 4–5. I cannot here discuss the issues she raises, in particular what kind of question it is whether the real can any longer be taken to be good.

40 Weil, "Human Personality," pp. 53, 71.

41 For a brief account of the context, see Gary L. Francione, *Animals, Property, and the Law* (Philadelphia: Temple University Press, 1995), pp. 179–82. Here is Francione's description of the particular scene: "In one scene, a lab worker poses with an animal who has massive cranial hemispherical sutures, and the lab staff, including Gennarelli [one of the directors], laugh at the animal, whom they tease as having the 'punk look.'"

42 See Barbara Harrisson, *Orang-Utan* (Singapore: Oxford University Press, 1987), p. 202; also Tom Harrisson, "Dear Cousin!" in Barbara Harrisson, *Orang-Utan*, pp. 9–26, at p. 25.

43 For a view of how we should treat animals, not open to the criticism I make here of animal rights theorists, see the writings of Stephen R. L. Clark, e.g., *Animals and Their Moral Standing* (London: Routledge, 1997).

44 See also Murdoch, *Sovereignty of Good*, p. 37; Murdoch notes, in discussing attention to reality in Weil's sense, that in such contexts "reality" works as a normative word.

45 See Sorabji, *Animal Minds and Human Morals*, p. 131, the reference to Sextus, and see Sorabji's entire discussion of *oikeiôsis* in relation to the possibility of extending the notion of justice to animals.

46 Sorabji, *Animal Minds and Human Morals*, pp. 131–32.

47 D. H. Lawrence, *Phoenix* (Harmondsworth: Penguin Books, 1980), pp. 344–45.

48 Leo Tolstoy, *What Then Must We Do?* (Oxford: Oxford University Press, 1946), pp. 99–100, 175.

49 Compare Hume (*Enquiry Concerning the Principles of Morals* [Oxford: Clarendon Press, 1927], pp. 190–91): we *can* impose whatever burdens we choose on animals, and the imposition of such burdens is to be limited by our feelings of kindness but not as a matter of justice. Justice is, for Hume, conceived in terms of property rights and rights that can be conceived in property-like terms. A system of justice is something we institute, including in it those who, independently of such a system, are capable of making us feel the effects of their resentment. (The *societas* linked to justice is that of rough equality of power.) Europeans believed, Hume says, that they could get away with treating Indians as not capable of effective response to ill treatment. He does not mention Africans, but his discussion implies that the enslavement of Africans is not injustice, and that the treatment of any enslaved people should be limited by kindness and humanity. Welfarist views about animals do not have to imply that enslavement of someone is no injustice if he cannot resist effectively. What welfarist views of animals have in common with Hume is the idea that there are some beings whose backs we do not *have to* get off; and that, given that we *are* going to exercise power over such beings, we should make the saddles as comfortable for those under them as is consistent with our remaining securely in them.

50 Bernard Williams, *Ethics and the Limits of Philosophy* (Cambridge, Mass: Harvard University Press, 1985), pp. 118–19.

9   Bioethics, Wisdom, and Expertise
Paul Johnston

The modern world is a world of experts. This concept of the expert has so expanded that nowadays people talk of experts in a wide range of areas, including ethics. This is an understandable development, but there is little doubt that Wittgenstein would have been appalled by it. He saw moral decision making as a highly personal matter and throughout his life opposed attempts to treat ethics as a science. His letters and conversations from the time of the *Tractatus* make clear his contempt for those who try to theorize about ethics, while the book itself denies that it is possible even to formulate meaningful ethical propositions. He expanded on this latter theme in his 1929 lecture on ethics, and in conversations with Moritz Schlick at around the same time he is emphatic that value judgments are beyond any kind of explanation. What is ethical, he claims, cannot be taught. "If I could explain the essence of the ethical only by means of a theory, then what is ethical would be of no value whatsoever." [1]

Twenty years later Wittgenstein's general philosophical position had changed significantly, but his hostility to moral theorizing remained. A regular theme of his conversations with O. K. Bouwsma was the harm philosophers have done in ethics, while the idea that there might be moral experts could still trigger a vigorous dismissive outburst. "He began talking about teaching ethics. Impossible! He regards teaching ethics as telling someone what he should do. But how can anyone counsel another? Imagine someone advising another who was in love and about to marry, and pointing out to him all the things he cannot do if he marries. The idiot! How can one know how these things are in another man's life?" [2]

The Notion of Moral Expertise

Although his reactions reflect Wittgenstein's personal views, they have also a philosophical basis, for the notion of moral expertise, by blurring the distinction between moral and empirical claims, reflects a failure to recognize what is really at stake in ethics. The essence of ethics is the claim that there are ways of acting that everyone should recognize as right and wrong. This claim cannot be derived from logic or from rationality, nor can it be demonstrated with (or supported by) empirical evidence. It is a claim about the correct way of understanding the world, but there is no independent way of resolving disputes about it. Different individuals will come to different conclusions about which view is correct, and, although this issue is of fundamental importance, there is no way of resolving the matter. On one side, there is the claim that the correct way of understanding the world involves recognizing certain ways of acting as correct. On the other side, there is the view that different judgments of human action simply reflect the different dispositions or different preferences of the people who make them.[3]

These may seem uncontroversial points, but they raise immediate difficulties for any claim to moral expertise. In particular, they highlight the fact that in relation to this kind of issue there is no guarantee that disagreements can be resolved. All claims in ethics can be questioned, from the claim that there are correct ways of acting to any specific claim about what those correct ways of acting are. Furthermore, it is not just that proof is impossible; rather, there is no scope for *any* independent adjudication of conflicting claims. This idea is more difficult to accept, for it is tempting to argue that while there can be no proof in ethics, it is still possible to show that certain sets of judgments are inconsistent or that certain judgments are incompatible with a particular principle. But this is misguided. It is logically possible to hold any combination of judgments on, for example, contraception, abortion, and infanticide; the point is that most people believe that many of the possible combinations are obviously wrong. Someone might claim that the human reproductive process should on no account be interfered with, but that there were situations where the product of that process was not sacrosanct. There is nothing inconsistent or incoherent in this position, but it is a position which virtually everyone in our society would agree is not the correct view of these issues.

As this example illustrates, claiming that a particular view is wrong is a substantive rather than a logical claim. It is one way of denying that it is the view that everyone ought to hold. Just as this type of intervention is substantive, so too are claims about what follows from a particular principle. We cannot say to someone, "If you believe x, you must believe y." Rather, what the person means when she says she believes x is shown by the specific judgments she makes about what is correct in different cases. For example, if someone says a doctor should always act in the best interests of her patient but should not give a patient forms of treatment she finds unacceptable, this does not show that her position is contradictory and confused. It simply shows that in her view acting in the best interests of the patient includes taking into account her beliefs about which forms of treatment are permissible. If we disagree with this view, we are effectively advocating a different principle, even if we too use the slogan, "A doctor should always act in her patient's best interests." The different interpretations of this slogan amount to different principles, and it would be senseless to ask which interpretation is correct. The real issue is which *principle* is correct.

Recognizing that there can be no independent adjudication of moral claims is crucial to understanding ethics. It also implies that there are fundamental differences between ethics and any science or academic discipline wherein an agreed-upon body of knowledge is gradually assembled through a process of debate and investigation. Unlike physics or medicine, ethics is a matter not of establishing facts, but of reaching a conclusion about how it is correct to act (and, as we have noted, it is quite possible for people to reject the idea that there are correct ways to act). Since there is no agreed-upon body of knowledge, there is no scope for a claim to expertise, for an expert is someone who claims authority on the basis of having mastered a body of knowledge. Against the background of an agreed approach, she tells us what we would know if we had had the opportunity to acquire the knowledge she has acquired. Where someone disagrees with an expert, the expert will usually be able to demonstrate the correctness of her view. Even where she cannot, the assumption built into the notion of an expert is that the person disagreeing would come round to the expert's opinion if she had the time and the opportunity to become more knowledgeable about the matters in question.

Trying to apply these ideas to ethics makes no sense. Take the example of an issue on which members of our society strongly disagree: abortion. Someone could certainly undertake a comprehensive survey of all the arguments that have recently (or not so recently) been advanced on abortion and could interview large numbers of people who have performed abortions, had abortions, or declined to have abortions, but none of this would make the person an expert on abortion. Her conclusions about whether (or in what circumstances) abortion was permissible would have exactly the same status as anyone else's. There would be no scope for her to claim the authority of an expert, for it is clear that other people could undertake the same investigation and yet reach very different conclusions. Unlike experts in other fields, this person could not say that her judgments were the judgments the nonexpert would make if only she had had more opportunity to study these matters. All she can say is: "I have studied these matters deeply, and these are the conclusions I have reached"; to which the nonexpert can reply, "I too have thought about these matters, and I am convinced that your conclusions are wrong."

Bioethics and Expertise

These arguments against the idea of moral expertise apply directly to bioethics. Given the nature of the questions it deals with, there is no basis for assuming that if we all studied bioethics, we would all come to the same conclusions. There is no guarantee that everyone would agree what the main issues were or which views were worthy of respect (or serious consideration) and which were not. This makes any talk of expertise fundamentally problematic. The bioethicist cannot claim to be the purveyor of a body of knowledge arrived at by an approach or method accepted by everyone, nor can she demonstrate the correctness of what she says. If she makes a claim someone disagrees with, there are no grounds for supposing that any period of time spent studying bioethics will lead the latter to change her position. As these points suggest, it would be wrong for the bioethicist to claim any authority in her interaction with the nonbioethicist. She has no basis for claiming a special right to initiate or guide the debate, and the status of her comments about the moral dilemmas of medicine is exactly the same as those of anyone else.

Perhaps most bioethicists would accept the above points and agree that it would be wrong for them to claim any special authority for their personal views. However, they are still likely to argue that there is a role for people like themselves whose professional training and familiarity with the moral dilemmas of medicine gives them the right background to help resolve the difficulties that arise in hospitals and elsewhere. They may also argue that in the absence of bioethical experts, some other group with no specific background in ethics will end up making the decisions. For example, issues about what sorts of experiments with human embryos are acceptable will be resolved by scientific researchers, while questions such as when it is right for a hospital to refuse treatment will be decided by doctors or hospital administrators who have not had any special preparation for such decisions. Since these sorts of decisions have to be made, is there not a role for properly trained individuals to assist in making them?

This is a valid question, but it takes us straight back to the issues we have just discussed, for what constitutes proper training in this area? And more fundamentally, what is the basis of the assistance this specially trained individual is supposed to be offering? One possibility is to see the bioethicist as an expert in moral argument. She is someone who has an understanding of valid argument in general and a familiarity with bioethical arguments in particular. This seems to define a relevant body of knowledge and so provide an appropriate basis for a claim to expertise. But this approach falls victim to precisely the difficulties we have described. Any claim the bioethicist makes about which arguments are valid or about whether a moral view is consistent or a general principle compatible with a particular type of action will be substantive. It will therefore involve claiming expertise in relation to judgments that are of exactly the same type as those in relation to which everyone agrees the bioethicist can claim no special authority. The problem here lies in the idea that someone might be an expert in moral argument. This idea fails to recognize that the difference between a good and a bad moral argument is not that the former is more in keeping with logic than the latter, but that it treats the issue in question in the way that we believe to be correct. It is a good argument because it makes the correct moral distinctions, includes everything we consider to be morally relevant, and, by assigning the correct moral weight to the various considerations, arrives at the correct answer. There is no essential difference, therefore, between claiming to be an expert in moral argument and claiming to be an expert in moral truth, for the only way of distinguishing between moral claims is by making an assessment of which we think is correct. Similarly, it would not make sense to claim to know all the relevant moral arguments in relation to a particular situation, for the number of ways of treating a moral situation is countless, while even to say that it raises moral issues is itself a substantive moral judgment.

To illustrate these points, let us consider a specific example. Suppose someone argued that we should refuse medical treatment to all pet owners because by spending money on their pets rather than on needy human beings they have shown themselves not to be deserving of assistance. Few people would take this argument seriously. Most would deny that there was anything wrong with keeping pets, would argue that there are a host of other relevant considerations (freedom of the individual, a doctor's duty to her patients, etc.), and would see the penalty as irrelevant or disproportionate to the supposed misdemeanor. Can the bioethicist therefore pronounce this argument bad? Can she say that anyone who advances it has ignored considerations that must be taken into account? Can she at least say that the position of someone who makes this claim is incomplete unless it includes a rebuttal of the obvious retort about freedom of the individual? The answer is no. There is no scope for expert pronouncements on any of these matters. Like anyone else, the bioethicist can condemn the above view, and she can explain why she believes that it is wrong. But she will not be able to show that it is wrong, nor would it make sense for her to claim that she was specially qualified to recognize what was wrong with it.

Against this, it may be argued that in some areas experience is important; for example, does not twenty years of studying the arguments surrounding euthanasia or a lifetime of working with the terminally ill give someone a special qualification for speaking about this issue? This seems a reasonable suggestion, and most of us would certainly be interested in what someone with this background might say. Nonetheless, it would be highly misleading to call this person an expert. Her judgments would have no special status; as with anyone else's account of how we should approach this issue, we would have to examine her arguments on their merits and decide whether or not we agreed with them. Furthermore, academic investigation or even frequent confrontations with a particular moral dilemma are no guarantee either of reaching the correct moral view on it or even of being the best person to state what the most important issues are. Years of study or daily encounters with one particular kind of situation may blunt or distort an individual's moral judgment rather than leading her to the correct view. Here, as elsewhere in ethics, there are no independent measures that can be used to avoid the need for substantive assessment. Whether we judge that an individual's studies and experiences have increased her wisdom or led her astray will depend on what we believe the correct views are.

Few writers on bioethics contradict these points directly, but there is a strong tendency to reject them implicitly by reintroducing the notion of expertise in a disguised form. Consider Zoloth-Dorfman and Rubin's comparison of the bioethicist to an experienced navigator:

> If the physician is the captain of the ship, then, we suggest, it is the ethicist who is the navigator; knowing the map, being familiar with the terrain and its complexities, calling attention to how the desired or expected course might be changed by the immediacy, temporality and particularity of a given case, and above all, guiding, but not controlling the course. The ethicist can offer direction, vision and even warning about the implications of a chosen path, contributing the unique perspective and tools of her profession. She can use questions, arguments, and moral suasion. But ultimately, unlike the captain, she cannot seize the ship's steering wheel. In this respect, although they may share some features in common, the roles of the captain and the navigator are inevitably distinct.[4]

This seems an admirably modest account, and it reflects the authors' explicit rejection of the idea that the bioethicist is a moral expert. Behind the apparent modesty, however, lie the same misguided claims. Doctors may be reassured to learn that the bioethicist will guide their actions rather than decide for them, but the former type of intervention just as much as the latter will embody a particular moral approach. The key point is that in relation to ethics there is no agreed-upon map that everyone recognizes; rather, different moral views offer different maps of life's dilem-

mas, and the bioethicist's acceptance of a particular map reflects her beliefs about what is important and morally relevant.

The comparison with a navigator is, therefore, misleading. There is no
basis for assuming that different bioethicists will agree in their assessment
of the terrain, and even if they did, their views would have no special authority. The bioethicist does not have a unique perspective to offer doctors, nor does she know something they do not. If she offers a direction, a
vision, or warnings, she can only do so in the same way as anyone else.
She and the doctor can discuss a moral dilemma like any other two human
beings; the most significant contribution philosophy can make to that
conversation is to point out the absurdity of any claims to expertise, be
they direct or indirect. The bioethicist may have thought about a particular difficult situation more than the doctor directly confronted by it, but
even if she has, she cannot give the doctor an independent expert assessment of it. All she can do is give *her* assessment; whether this contains
more wisdom than the assessment of the hospital janitor will be a matter
of judgment and may vary from case to case. On life's ship, any of the pas-
sengers may issue warnings about the dangers of the course, but there is
no expert navigator whose professional studies can enable her to pronounce authoritatively on where the dangers really lie.

Bioethics and Wisdom

The above arguments are based on a recognition that there can be no independent adjudication of moral claims. However, nothing I have said implies that moral judgments are merely a matter of opinion or simply express the preferences of the person who makes them. On the contrary,
ethics is defined by the claim that there are correct ways of acting and
therefore correct and incorrect judgments of human action. It is those
who reject ethics who believe that judgments of human action can only
ever reflect the preferences or dispositions of an individual or a group of
individuals. This is a coherent claim about how the world should be
understood, but it is the opposite view to that held by the moralist. She
denies that all approaches to human action are on a par and holds that
some approaches are wise and others foolish, misguided, or even evil. According to the moralist, the correct way of understanding the world includes having the correct view about how people should act—indeed, for

her it is this, rather than mere knowledge, that differentiates the wise from the foolish. Someone who believes that the earth is flat is mistaken, but someone who believes that there is no such thing as objective right and wrong has misunderstood the world in a much more serious way.

As this suggests, wisdom and knowledge are very different concepts. One obvious difference is that no course of study will necessarily bring wisdom. A more fundamental difference is that there are no independent ways in which wisdom can be judged. We can test someone's scientific knowledge by subjecting her claims to appropriate empirical investigation, but we cannot do this with wisdom. The only way to assess someone's wisdom is by assessing the actual judgments she makes; all we can do is to reflect on what she says and decide whether we think it embodies the correct approach to the issue in question. Of course, if we come to believe that someone is wise, we may accept a judgment she makes even if we have difficulty understanding why it is right. Doing this, however, would involve being so convinced by this person's moral insight that we were ready to take a particular judgment on trust. Furthermore, many people today would argue that it is wrong ever to put one's conscience in another person's hands in this way. Be that as it may, the general point is that we *can* give someone a special role or even authority in this area, but when we do, it is not on the basis that the other person possesses knowledge or information we lack but because we judge them to be wise.

These points about wisdom underline the fact that all moral claims are substantive, but they are not particularly helpful in terms of providing a basis for bioethics. A bioethicist could base her interventions on a claim to wisdom rather than expertise, but then she would be claiming to know better than other people what is right, and that is not a claim people in our society are encouraged to make. Let us, therefore, explore a different idea and see bioethics as involving not wisdom but the mastery of a technique. On this approach, the bioethicist would be an expert in helping people develop their powers of moral reflection. The difficulty here is that helping someone with her moral thinking inevitably affects that thinking. Once again, therefore, the notion of expertise is inappropriate. The bioethicist cannot say that using drugs that have been tested on animals is something a doctor needs to think deeply about, but that using drugs produced from genetically modified animals is not. Similarly, it would be wrong to think that a bioethicist could offer expert help to a doctor simply by confronting the doctor with the implications of the doctor's own views, for, as we noted earlier, a principle and its implications define each other, so the only person who has any authority in detailing the implications of a principle is the person who advances it.

Some of the comments made by Zoloth-Dorfman and Rubin suggest that they favor this kind of approach to bioethics. For example, they emphasize that bioethics is a conversation: the bioethicist raises questions and through discussion helps focus the doctor's deliberations. Once again, however, they imply that the bioethicist has a special status, for they suggest that she lead the discussion. This is confused. Any guidance the bioethicist offers will reflect particular substantive views, and no amount of professional training can give those views a special status. The suggestion that the bioethicist guide rather than determine, question rather than advise, masks but does not undermine this point, for claiming to know the right questions involves the same kind of claim to moral insight as claiming to know the right answers. For example, if someone believes that there are five criteria that must be met before it is right for a doctor to help a patient die, then it is partly a matter of style whether she states this directly or raises a host of questions that suggest that these are the considerations that ought to be decisive. Thus, while there can be no objection to the idea that bioethics is a conversation, the suggestion that it is not a conversation between equals is confused and misleading. We are back to the misguided idea that the bioethicist is in some sense a moral expert.

Rejecting the idea that the bioethicist is an expert does not rule out the scope for bioethics; rather, it raises the question of the role bioethics plays in our society. Why is it that a professional discipline of this kind has emerged? Part of the answer is that we live in societies where no one moral code is dominant, and where there are no people who are universally accepted as wise and have the role of resolving difficult moral issues. As a society, however, we have to make decisions about what we will prohibit and what we will permit. For example, we have to decide which kinds of experiments with embryos our laws will allow, or how we will treat people in a persistent vegetative state. The process by which we make these decisions is neither simple nor clearly defined. There is a democratic aspect to this, insofar as a decision that was strongly in conflict with what most people think would be unsustainable, but this is not the sort of issue that is constantly being taken to the ballot box. Rather, these matters are dealt with in a more complicated way, with political institutions, the courts, and other bodies interacting to determine the prevailing view at any one point in time.

This is the context within which bioethics operates. Our society's debate about medical ethics involves not just practitioners, lawyers, judges, politicians, and "ordinary" people, but also a group of individuals whose academic qualifications and activities relate specifically to these issues. The role these people play in the debate is twofold. On the one hand, they contribute to the debate by arguing in favor of particular approaches or by criticizing the reasonableness of proposed distinctions. On the other hand, they record the debate and the different views that are being put forward. Ironically, there is scope for a claim to expertise here, but not moral expertise. If bioethicists are experts, they are experts on the state of play in the bioethics community. While the bioethicist has no authority to state the correct view, she is qualified to explain the main approaches to a particular issue that are currently considered reasonable in our society. Of course, there is no guarantee that any of those approaches is correct; someone may claim that they are all fundamentally flawed and that society (or the elites that play a primary role in determining these issues) is misguided or depraved. But that simply underlines my earlier point that in ethics there is no independent (or unchallengeable) way of determining the correct view.

Conclusion

As far as Wittgenstein himself is concerned, it seems likely that he would have been hostile to the idea of bioethics as a profession. He certainly seems to have thought that the intellectual climate of academia was more likely to foster moral confusion than wisdom, and he would probably have had greater faith in the moral instincts of a doctor who had never had any bioethical training than in those of one who had. In philosophy he saw cleverness as a temptation, and for writers on ethics this danger is even greater. The professional moral thinker risks seeking academic success or opportunities to demonstrate the power of her intellect when the real issue is trying to find the right way to live. Wittgenstein worried that studying philosophy would lead his pupils away from wisdom rather than toward it, and he would almost certainly have had the same worry in relation to bioethics. If a doctor had a choice between reading Tolstoy's version of the New Testament or reading a learned article on whether turning off a ventilator involved withdrawing treatment or withholding it, Wittgenstein would probably have recommended the former.

All of this, however, is speculation about the personal views of one man. From a philosophical point of view, what is important is clarity, and the aim of this essay has not been to attack bioethics but to reject the idea that there can be moral experts. There is no scope within ethics for a claim to this kind of authority. On the contrary, in the face of the dilemmas of medicine (or any other moral dilemma) we are all in the same situation. As Wittgenstein noted, "[A]t the end of my lecture on ethics, I spoke in the first person. I believe that is quite essential. Here nothing can be established. I can only appear as a person speaking for myself." [5] For the professional bioethicist, *not* appearing as an expert may be a struggle, particularly as other members of society may want to push their own responsibility onto the shoulders of an "expert." The point, however, is not that the bioethicist should avoid taking a stance, but that she should be clear that she does so as an individual and not as an expert. Like anyone else, she can claim to have insights into a particular dilemma (or can advance claims about what the key issues are or about what responses are reasonable), but any such claims are claims to wisdom, not expressions of expertise.

Notes

1 Friedrich Waismann, *Wittgenstein and the Vienna Circle: Conversations*, ed. Brian McGuinness, trans. Joachim Schulte and Brian McGuinness (Oxford: Blackwell, 1979), pp. 116–17.

2 O. K. Bouwsma, *Wittgenstein: Conversations, 1949–1951* (Indianapolis: Hackett, 1986), p. 45.

3 For a fuller discussion, see Paul Johnston, *Wittgenstein and Moral Philosophy* (London: Routledge, 1989), and Paul Johnston, *The Contradictions of Modern Moral Philosophy: Ethics after Wittgenstein* (London: Routledge, 1999).

4 Laurie Zoloth-Dorfman and Susan Rubin, "Navigators and Captains: Expertise in Clinical Consultation," *Theoretical Medicine* 18 (Dec. 1997): 4.

5 Waismann, *Wittgenstein and the Vienna Circle*, p. 117.

10 Wittgensteinian Lessons
on Moral Particularism
Margaret Olivia Little

If anthologies and conference panels are any indication, bioethics has lately been greatly exercised over the role of particularity in moral judgment. Discussions about moral judgment circle back, time and again, to the roles of detail and generalization, of particular and principle. The issue is of interest to moral philosophers generally, of course, but it takes on a certain exigency and taps into a special set of crosscurrents for those who do bioethics. After all, bioethics as a discipline was born from the conviction that the messy and urgent moral dilemmas found in health care matters could benefit from systematic ethical theory; yet it is a discipline that addresses itself, in large part, to people whose own professional experience underscores at every turn the importance of attending to detail—in medical diagnosis, for example, where crude application of generalizations can prove fatal, or in caretaking, where decisions about how best to provide a given patient with a dignified death, say, or gentle respect seem to be outside the reach of any algorithm. Then again, bioethics is also about public policy, where injunctions to attend to the particularity of each case run counter to the whole point of setting what must perforce be generalization over instances. And so it goes, leaving those reflecting on method in bioethics to debate how best to capture the role of particularity in moral decision making.

An enduring figure in such conversations—some might say an enduring threat—is what we might call the radical moral particularist. Such a theorist makes the rather dramatic claim that moral answers cannot be captured in general formulae. She claims not simply that we should be properly attentive to the relevant details of situations before we can apply any rule or principle to them, but that there *are* no rules or principles, even enormously complicated ones, capable of codifying the moral landscape. Thus, while all agree (or would, if they bothered to reflect on the matter) that it is crucial to attend to nuance and contextual detail when thinking through moral matters, the radical particularist makes the further claim that, as it were, it's nuance all the way down.

While some sign on to the particularist's bandwagon with enthusiasm (born occasionally, no doubt, from naïveté about the party platform), it is fair to say that most theorists feel deep resistance to the doctrine. In some cases, this resistance is grounded in a sort of philosophical annoyance, reflecting the conviction that most particularists forget to provide any argument for their position, contenting themselves with simply announcing their pessimism about the existence of adequate principles or with constructing ever more imaginative counterexamples to proposed principles, while failing to see that such counterexamples, even if accepted, might simply indicate that adequate principles are complex rather than nonexistent.

Usually, though, it is something much deeper than mere frustration grounding the resistance that so commonly greets particularism. The persistence with which those increasingly imaginative counterexamples are met with efforts to render principles ever more nuanced, after all, suggests an optimism that mirrors the particularist's pessimism; and such a presumption does not arise *ex nihilo*. In fact, the presumption that there are moral principles capable of codifying morality is deeply tied to the presumption that there had *better* be such principles. Resistance to particularism, in other words, is often underwritten by the worry that something deeply important would be lost if particularism turned out to be true.

The worry finds expression in various forms. Most prosaically, it shows up as bewilderment about how, under particularism, we are supposed to proceed when we are unsure or disagree about moral matters. After all, according to our most recent tradition, the whole point of ethical inquiry is to provide a *method* or *system* for providing moral answers—a view especially congenial to bioethics, which most often regards its mission as providing advice, in the form of theoretical order, to those working in the trenches of clinical practice or public policy. But moral particularism, it is felt, abandons us to navigate the thicket of our confused intuitions armed with nothing more than the injunction to "make wise judgments."

Pressed more deeply, the worry expresses a set of philosophical concerns that particularism undermines the status of morality. For instance, many charge that particularism is incompatible with the very notion of justification. Justification, it is urged, proceeds by subsumption under generality. If there are no codifiable generalities governing individual cases, then not only have we no method for finding answers, but nothing can function as a reason to be offered in defense of any conclusions we do reach. Or again, particularism is felt to cast doubt on the pretensions— already precarious—that morality is an objective enterprise. If the particularist is right, it is sensed, morality would finally be confirmed as something disastrously idiosyncratic—a hard blow to those who search valiantly for answers to heartfelt dilemmas. In one form or another, then, resistance to particularism can often be traced to feelings of philosophical vertigo, to a sense that the doctrine pulls the rug out from under morality.

Mention of vertigo brings to mind Stanley Cavell's wonderful passage about Wittgenstein's views on the grounding of normative authority.[1] In Wittgenstein's later philosophy, what it is for something to count as the right move—what it is to follow a rule—is located in the "whirl of organism" or "form of life" rather than anything transcendent of our practices. Contemplation of that idea, Cavell states, can give rise to a sense of terror, a dizziness that the appropriateness of moves rests on thin air. But, as Cavell points out, the antidote to this feeling, according to Wittgenstein, is not to redouble our efforts at finding a metaphysical grounding, but to realize that we were never standing on anything more than our practices—and yet we were standing. The solution to the vertigo, then, is to reveal and disabuse ourselves of the problematic philosophical models— models so familiar we often forget that they are, nonetheless, pieces of theory—that are responsible for the anxiety. I want to suggest that something analogous can be said about what drives much of the widespread resistance to moral particularism. For notwithstanding the claims of those who find particularism *ad hoc*, the particularist's doubt does not stem from philosophical obsession with counterexamples or lazy extrapolation from them; it is not brute pessimism floating free of other philosophical commitment. The particularist doubt is a natural outgrowth of a certain picture of morality, and Wittgenstein's later philosophy offers lessons that should help dispel the sense that the picture is anything suspicious or alarming. The lessons I have in mind span a range of Wittgensteinian themes, from his philosophy of mind and epistemology to his rejection of "metaphysics"—indeed, just about everything but his ethics, which in this essay are cheerfully laid to one side.[2]

In what follows, I lay out in rather cursory fashion the picture or model of morality that gives rise to a particularist approach to ethics and outline how certain lessons from Wittgenstein might cast reassuring light on its implications. The result is hardly a defense of moral particularism (which, after all, might be rejected for reasons quite separate from loyalty to models Wittgenstein would dispute). Reflection on Wittgensteinian themes, though, can at least help to highlight that many concerns about particularism depend on acceptance of certain contestable models. It might even be a first step toward curing the dependency.

A deeply influential theme in the last few centuries of moral philosophy holds that ethical inquiry is the search for the architecture of morality. It is a search for the patterns that lie behind specific cases; more than that, it is the search for generalizations that hold not as mere accidents (as when we realize that Dora is usually cruel to Jack), but with a level of necessity, revealing the very nature or structure of the moral landscape. Just as scientists try to parse out how the forces of physics interact systematically, moral theorists try to capture how moral considerations so identified (such as the requirements of justice and beneficence) are ordered in relation to each other. And again, just as scientists work to unearth laws linking, say, the property of temperature to the property of mean kinetic molecular energy, the job of moral theorists is to identify which natural properties make an action count as just or beneficent.

Moral particularists have a very different picture of morality. Without questioning that there are moral answers, they question whether those answers are, or need to be, backed by any architecture, whether they constitute pieces that fit together to form a structure with recoverable lawlike relations. To be sure, no one (sensibly) rejects principles that tell us to "respect autonomy" or to "be just." But the particularist denies that we can unpack those very abstract principles into generalizations that are both accurate and contentful enough to guide action. Particularism denies that we can codify how the demands of justice and those of autonomy weigh up, or which nonmoral features suffice to make an action just in the first place.[3]

The core intuition driving this picture is that moral properties are "shapeless" with respect to the nonmoral.[4] There is no way of marking out the boundaries of moral concepts in purely nonmoral terms; items grouped together under a moral classification such as "cruel" do not form a kind recognizable as such at the natural level. Of the infinitely many ways of being cruel—kicking a dog, teasing a sensitive person, and forgetting to invite someone to a party might each qualify—there is no saying what they have in common (and why, say, the pain inflicted during a spinal tap is different) except by helping oneself to the moral concept of cruelty. More deeply, whether those examples themselves qualify as cruel depends irreducibly on the contexts in which they are situated. A set of features that in one context makes an action cruel can in another carry no such import; the addition of another detail can change the meaning of the whole. Intentionally telling a falsehood is often a breach of fidelity, but it need not be: imagine a prisoner lying to a sadistic Nazi guard, or a parent lying to her child while playing the game Cheat.[5] Indeed, the very "valence" of a feature's contribution is context dependent. The fact that an action is fun, to give an example cited by Jonathan Dancy, often counts in its moral favor but at times may be precisely what makes it morally problematic: that the sadist enjoys inflicting pain seems precisely what is wrong with the action, and not the "moral silver lining" of the situation.[6]

The point is not to deny that natural features serve as "good"- and "bad-making" properties. When classifying an action as cruel or just, we certainly regard the moral status as obtaining "in virtue" of certain of its nonmoral features: those natural features are what *make* the action cruel, are the *reason* it is kind. The point, rather, is to deny that such considerations carry their reason-giving force *atomistically*.[7] Natural features do not always ground the same moral import, which then goes in the hopper to be weighed against whatever other independent factors happen to be present. The moral contribution they make on each occasion is holistically determined: it is itself dependent, in a way that escapes articulation, on what other nonmoral features are present or absent. It is not just that we have not bothered to fill in the background conditions because they are so complex, mind you; holism is not complicated atomism. The claim, rather, is that there is no cashing out propositionally, once and for all, the context on which the moral meaning depends. To apply a nice example of David McNaughton's, natural features carry their contribution to an action's moral status in the way that a given dab of paint on the canvas carries its contribution to the aesthetic status of a painting: the bold stroke of red that helps balance one painting would be the ruin of another, and there is no way to specify in nonaesthetic terms the conditions in which it will help and the conditions in which it will detract.[8] Just so, whether a given feature counts as any moral reason at all—and if so, with what valence—is itself irreducibly dependent on the background context. While the moral properties of actions, then, are in some sense determined by their natural features, there is no pattern discernible outside the evaluative practice to how those individual determinations add up. A culinary example might help illustrate. The tastiness of a spaghetti sauce is surely determined in some sense by the chemical properties of its ingredients; but the contours of categories like "tasty" cannot be seen at the chemical level. There is no writing out once and for all a checklist of possible ingredients one might put into a good spaghetti sauce, and no specifying the conditions under which a given ingredient will add and when it will detract from the sauce (fennel bulb can be a terrific addition, but not if one has already added tarragon, unless perhaps one has thrown in some romano cheese to em-phasize sharp tones, and so on). A chemist who analyzed a sample of deli-cious sauces in her lab but who lacked taste buds—who failed to see the "gustatory point" of the sauces—would not be someone to trust in the kitchen. She would be unable to judge of new ingredients which are worth trying and which are better left on the shelf, unable to judge when famil-iar standbys should not be deployed. Just so, someone who remains an outsider to the evaluative point of moral categories will not learn "how to go on." However many cases of cruelty one is shown, reflection on their natural properties will not afford patterns that will enable one to catch on to what cruelty really is. Novel instances will be missed, and others will be falsely classified as cruel because of similarities that turn out to be superfi-cial. The particularist begins by rejecting attempts to codify relations be-tween nonmoral and moral properties. The resultant picture also leads to a rejection of efforts to systematize relations among moral properties. Mo-rality, unlike, say, arithmetic, is not an internally codifiable system. Hav-ing identified the substantive moral considerations a situation presents—that there is, say, a duty of beneficence to perform an action but a duty of justice to avoid it—there is no algorithm for determining whether the action is morally right. There is no saying once and for all when justice trumps mercy or mercy justice: whether to give punishment or clemency turns on the details of the case. How such conflicts are to be resolved depends, in a way that escapes articulation, on the context in which they occur.

Such a stance, which has been advocated by many, makes sense given the picture sketched above. Different instances of each type of moral consideration can stand in very different weighing relations to one another; for there are different ways of being a demand of beneficence or justice, and how the elements of an individual case form here to constitute kindness influences how they stack up to the elements that here constitute justice. To give an analogy with color, verdicts about when colors clash and when they complement one another are not ones that can be made at the generic level of red, green, blue, and so forth, for the simple reason that different species within these genuses can, even in similar contexts, stand in very different aesthetic relations to one another. Khaki and fire-engine red might be a smashing choice for the exterior of my house, while chartreuse and mauve would be (trust me) really bad. Just as one cannot infer which colors to put together in a particular instance if one only knows what color-genus they fall into, one cannot infer how to resolve a particular moral conflict if one only knows what generic types of moral considerations are involved. Certain cases of beneficence should give way to certain instances of justice, while others should hold sway over them. Notice now that this picture backing particularism simply obviates what is perhaps the most persistent traditional argument offered in defense of deductive principles.[9] It is often said that if a moral conclusion is to count as the right one, such principles *must* be lurking in the background. After all, it is thought, if the reason for the conclusion is adequate—if it really operates as a reason—it must be an instance of a generality that holds through all circumstances: an exception somewhere shows that one must refine one's claim about what actually counts as the reason. It should be obvious at this point, though, that such an argument has tacitly assumed that moral considerations or reasons function atomistically. If, as the particularist suggests, they instead function holistically, a set of considerations can function as a reason here—truly function, and not simply be an incomplete rendition of a reason—and yet not count as such a reason in another context. The properties that here make an action cruel can else-where, in the presence of additional features, help make it kind; or again, the properties that are here morally irrelevant can be rendered salient when accompanied by other features. If we feel the urge to incorporate the background context into the reason, we betray an alliance to the view that reasons, in order to function as reasons, must everywhere function the same way—precisely what the holist rejects.

For many, the idea of such holism, and the rejection of an architectural model of morality that follows from it, induces a distinct sense of philo-sophical queasiness. That queasiness has a variety of sources, some of which get (energetically) articulated, some of which tend to operate sub-terraneously.

A leading worry is that particularism renders our moral categorizations unacceptably subjective. If the particularist is right, someone who gazes upon the natural world will find that certain things get clustered together by us as "cruel" and others are excluded, but according to no pattern that can be seen at that natural level. How, then, is there any measure of consis-tency? How can we speak of the criteria for categorizing something as cruel as shared and not idiosyncratic? Indeed, how could we ever come to understand what is meant by, say, "respecting patient autonomy" in this picture? These puzzles underwrite a sense that if the particularist is right, our moral categorizations cannot be said to be responding to anything in the world; morality turns out to be a matter of taste, and illustrating the doctrine of holism with examples from art—not to mention spaghetti sauce—simply confirms it.

These intuitions, though, reflect a cluster of biases. First and foremost, it is a falsely narrow notion of consistency that counts us as going on in the same way only if the measure of sameness can be found at the natural level.[10] Such a notion privileges a subset of our commitments, namely those from science, as the judge of all others. But this is to engage in meta-physics in the pejorative or Wittgensteinian sense of the word: it is allowing a substantive preconception of what facts and objectivity look like to constrain ahead of time our view of what the world contains. (It is one thing to say, if we do, that science is objective, and quite another to say that science is the exclusive arbiter of objectivity.) The fact that the category of cruelty has no shape at the natural level does not mean it has no objective shape: it has the shape, precisely, of cruelty. Likewise, the proper measure of consistency is that one calls cruel those things, and only those things, that are cruel. It is of course possible that there is in fact no such property, and in this case it will turn out that there is no objective measure by which our categorizations count as consistent. The point, though, is that this must be settled by open-minded investigation, with terms appropriate to the subject matter. It is something that must be settled, that is, by rolling up our sleeves and seeing if the worldview supported by our best evidence turns out, on reflection, to include evaluative properties such as cruelty. It is not to be settled by appeal to some philosophically driven picture of what facts and proof must look like, not by ruling out of court the possibility of such properties because, against the paradigm of science, they look, in John Mackie's phrase, "queer." [11] Just so long as our commitments to moral properties survive reflective scrutiny, we have available a criterion of consistency in our moral categorizations. Consistency, then, need not be found at the natural level in order to be present. Of course, the particularist claims not just that the conditions of something being cruel cannot be spelled out in natural terms, but that they cannot, except trivially, be spelled out at all. Against one traditional philosophy of mind, this will seem worrisome: how can we explain what leads us to count certain acts but not others as cruel, and in a way that points to anything shared, if we cannot articulate the criteria of application?

But the model is just that—a model—and it is one that Wittgenstein urges us to reject. One of his central points is that we can share things, such as understandings, skills, and practices, that outstrip finite sets of propositions. If we feel that such a picture leaves an unsettling lacuna in our psychological explanations, we are reminded that it is fantasy to regard the lacuna as filled where codification is available. After all, even with mathematical rules such as "add two," our previous behavior is consistent with an infinite number of rules; the idea that we possess something, such as a Platonic concept, that guarantees for us the trajectory of moves is an illusion. In the end, we can explain why we make the moves we appropriately do only by saying that we share a practice. [12]

If we start to wonder how someone could ever come to catch on to a rule whose shape cannot be cashed out, we should remind ourselves that this question is generic as well. How, Wittgenstein asks, can we ever come to catch on to a given rule from a finite set of examples, given that the examples we are shown could logically have been the products of an infinite number of rules? His answer, of course, is not to retreat into skepticism, but to emphasize that we can, as members in a "whirl of organism," outstrip the conditions of learning. Thus it is certainly true that we will come to understand a moral concept such as fidelity by reference to certain paradigmatic examples, such as intentionally told falsehoods. But this is just to say that we learned to become competent with the concept under circumstances in which the most easily accessible breaches of fidelity happened to be actual or mythic cases of intentional falsehoods. Once we have come to catch on to the concept, we are able to discern the very different natural shape fidelity and its breaches take in different contexts. To think we cannot is to confuse the conditions of learning with the content of what is learned—a conflation Wittgenstein is at pains to warn against.[13] The conflation, to be sure, is a tempting one, for much of the reason it can seem so natural to model moral competency as the possession of codifiable generalizations is that appealing to such generalizations is often such a helpful pedagogical move. Moral generalizations or principles, that is, are often invaluable devices in getting people to start understanding what is meant by "cruelty," say, or "kindness." But one should not confuse the *usefulness* of a generalization with its *truth*. In point of fact, the pedagogical usefulness of such generalizations is entirely parasitic on the fact that they are grossly oversimplified—indeed, in a word, false. As with any skill (driving a car, cooking, interpreting patients' expressions of pain), a novice in moral understanding needs to be given intentionally crude rules that bracket exceptions and keep complexity off the radar screen in order to guide the practicing that is necessary to increasing expertise. (Think of how necessary it is at one stage to tell a child, "One should never, never run across the street!" even though one ought, of course, to do just that in any number of situations, such as running from a tornado.) So, too, one might well introduce someone to the concept of fidelity by teaching her that one should never lie. But it simply does not follow that intentional falsehoods are always wrong or even wrong-making. It is just a fundamentally misguided picture of pedagogy to think that one is always giving an *analysis* of a concept when one is helping someone catch on to its meaning (a picture that is, nonetheless, a mainstay of overintellectualized parents, who can be seen spluttering to find and convey the necessary and sufficient conditions of some complicated concept such as kindness to a two-year-old, when what is really required is provision of a false universal and confidence that only experience can—and will—do the rest).[14]

The doctrine of holism, then, need represent nothing untoward. Indeed, if we need any further reassurance, we can simply look to mainstream theories of epistemology. I have often thought it odd that so many find the idea of holism downright disorienting when discussing moral reasons and yet accept it as near-dogma when discussing how epistemic reasons operate. One of the most enduring of Wittgenstein's epistemological insights in *On Certainty* (an insight taken up by Quine and others who "naturalize" epistemology) is that beliefs and experiences do not carry their justificatory import atomistically.[15] The fact that one has a perceptual experience of seeing a table can be excellent reason to conclude that there is such a table, but in other contexts it will count as excellent reason *against* drawing such a conclusion—as when you have just helped yourself to a hefty dose of a psychotropic drug that has in the past given rise to hallucinations of tables. There is no way to codify the conditions under which an experience of seeing a table is evidence for there being a table. Evidential justification is an irreducibly contextual enterprise, holistically dependent on the background of beliefs we bring to a given epistemic situation. But clearly, though evidential justification functions holistically, we can still evidently learn what counts as a good epistemic reason and what does not. Holism, then, is a completely familiar model, not just of subjective-sounding realms such as aesthetics, but of realms as indispensable as epistemology.

The most famous objections levied against moral particularism, of course, are epistemological in nature. Particularists have famously emphasized the possibility of coming to moral knowledge not by invoking generalizations that allow us to infer moral conclusions from such details, but by seeing what moral properties such details together ground.[16] To many, this immediately raises the specter of moral intuitionism. The particularist, it is charged, is guilty of positing some odd faculty that mysteriously allows us to pick up on moral properties. Such a view, it is claimed, provides us no account of how we might move from confusion to insight or from disagreement to accord; indeed, it provides no way to tell who has the special faculty except by the dubious (not to mention political) measure of crediting those whose verdicts agree with one's own.

But talk of moral discernment need involve nothing mysterious, *outré*, or *ad hoc*. We would explain the ability to apprehend that something is cruel or some action called for in broadly the same way we explain our ability to apprehend that something is a table: not by appeal to any special sense organ, but by appeal to a much more familiar faculty, namely, the capacity to apply a concept appropriately or, as Wittgenstein would put it, the ability to follow a rule.[17] Our "sensitivity" for moral properties would not be the sensitivity provided by some modular sense organ. Rather, if we can be said to apprehend that something falls under the classification "cruel," it would be by attending to the complexities of the case, discerning what is salient, making appropriate discriminations, and employing our understanding.

We earn the right to claim *apprehension*, of course, only as long as we earn the right to talk of moral properties and truths. But, once again, the crucial point is that such commitments are not to be denied *ex ante* of our investigations and reflections—are not to be decided ahead of actually "looking and seeing," as Wittgenstein would say, how to describe our best commitments. The temptation we must resist is starting out with a substantive and narrowed notion of what "epistemic access" to the world looks like and using it as leverage against the sorts of truths we should acknowledge. Our picture of what the world is like and our picture of what our methods of access to it are like must arise in tandem. Discernment, if we have it, then, is a matter of knowing how to go on with moral categories. Put in updated Aristotelian terms that are congenial to Wittgenstein's later philosophy, the view might best be thought of as a skills model of moral discernment. To say that moral discernment is a skill is to signal that it cannot adequately be modeled as the possession of algorithms: knowing how, as the saying goes, cannot be reduced to knowing that. (Again, this does not mean that it is rawly subjective or idiosyncratic, for we can share standards of right and wrong moves that cannot be reduced to finite sets of propositions.) Like any skill—playing chess, driving, diagnosing disease, parenting, interpreting X-rays—the skill of moral discernment takes development. One begins as a novice, trying to catch on to the rudiments; moves with practice to journeyman, in which one is roughly competent but far from agile; and, with enough experience and help, arrives at the smooth and intricate, though never perfect, ease of the expert. Developing this skill is no easy matter. Moral categories such as "kind" are far harder to grasp than are categories such as "chair"; it takes experience and subtlety to understand the difference between being kind and being nice, between false and genuine charity, between tough love and abandonment. Invocation of moral principles, once again, can be enormously helpful in training someone into this competency—one can get a leg up on the notion of kindness by starting with the principle always to be nice. But the usefulness of such principles to novices does not mean reliance on them is the model for experts. With moral wisdom, as with any skill, it is a sign of maturity to be able to let go of the guidebooks, cookbooks, and primers, and to exercise directly one's ability to judge.

Particularism has often been charged with leaving us speechless in the face of disagreement. But to say there is no architecture one can invoke when arguing one's case does not mean there is nothing to say. Sometimes it is a matter of drawing attention to salient details others have missed: we are obtuse creatures, often unaware of what is right in front of us (think how often people fail to see the pain of their own family members). Sometimes it is a matter of articulating others' underlying biases. Other times, seeing more clearly is a matter of gestalting the individual elements one already knows in a way that permits recognition of the further properties those elements fix, as when we gestalt the dots of a pointillist painting and see it as a still life. Once again, moral rules or generalizations can be important to these enterprises. Alluding to moral principles such as "inflicting pain is cruel" can serve as a heuristic, helping to highlight what is, in the given instance, a salient moral feature. But again, this is just to say that the principle is enormously helpful, not that its generality is true. A heuristic, after all, is a technique for getting someone to see or interpret a situation in a certain way, akin to turning one's head sideways to see the duck in the duck-rabbit figure—something that, to work, need not be a proposition, much less a true one.

One may feel the urge to protest that such methods may all be deployed and yet agreement not be reached. To be sure, none of these measures is guaranteed to work. But as Wittgenstein emphasized, it is an illusion to think there is *any* kind of argumentation or justification—even that offered by deductive proof—that somehow ensures that all will understand: reasons are not made to convince the universe. Rather, understanding the authority of a given sort of reason, being amenable to its force, requires that one be inside the particular practice that gives the reasons their home and their life. One will not grasp the point of certain ethical considerations, or the force of certain maneuvers in asking for or giving reasons for moral conclusions, if one is too distant from the evaluative "way of life" that anchors them. But this is as true of logic as of ethics. And, of course, those well ensconced within a practice are not thereby always perfectly reasonable by its own lights: reasons do not *force* compliance, they normatively require it, and holistically define a boundary such that radical noncompliance indicates one is no longer playing the game.

Of course, not all practices are ones that yield truths. Part of what defines a practice is the repertoire of reasons it contains for assessing moves and arbitrating disagreements. If the resources contained within a practice yield reasons that are too impoverished, thin, or conflicting to reach a modicum of consensus, then we have found grounds for concluding that the discourse does not, in fact, describe the world; it is a game that looked like a describing game but is in fact something else. If it turns out that the moves one can make in explaining and defending claims about what counts as cruel are too thin and indeterminative, we have reason to abandon the idea that there is any such property "cruelty" that things can be. This, for Wittgenstein, would be the fair test for whether or not to accept that a realm gets to talk of truth.

Now whether morality passes or fails this sort of test is a familiarly complicated and contested matter. The point I want to underscore here is that, following Wittgenstein, we should not pretend that the matter turns on whether morality can be codified. Amenability to codification does not settle amenability of rational agreement, for the question will recur of what resources the practice has available—rich or thin—for arbitrating amongst disagreements over the principles. And resistance to codification does not preclude such objectivity: an irreducibly skills-oriented practice can be one that admits of a large degree of consensus. Once again, I am setting to one side the question of whether morality admits of a sufficient such propensity for consensus (I will stop only to caution the reader from too quickly assuming, from superficial disagreements, that it does not). The central point is that repertoires of reasons can be too thin and indeterminative for realms that are principle governed and too rich and robustly guiding for areas that are not.

Notice, finally, that the complexity and familiarity of the components that comprise the skill of moral discernment means that the theory, unlike classical moral intuitionism, has the potential to provide a substantive account of moral error and disagreement. If apprehending things like cruelty involves exercising a variety of general abilities and sensitivities —being aware of what is salient, drawing relevant discriminations, remaining undistracted by irrelevancies, and of course, understanding the moral categories at issue—we can assess whether these are present by means relatively independent of what moral verdicts they are delivering on a given occasion. This is not to say we can assess the presence of the skill in complete independence of our general moral outlook; but that is a familiar lesson of Wittgensteinian epistemology. There is no transcendent viewpoint from which to judge the reliability of our methods *tout court*.

If we take care to work with a notion of discernment unburdened by associations with classical moral intuitionism, worries should abate that the particularist's notion of moral discernment represents anything that is in principle problematic. For many, though, the real stumbling block to accepting particularism lies not in suspicion of moral discernment, but in suspicion that discernment is the only thing particularism has up its epistemic sleeve. Particularism, it is often thought, implies that the only way we can get justified moral beliefs is to exercise discernment on a case at hand. After all, its central claim is that there is no cashing out when inflicting pain, say, constitutes cruelty and when it constitutes kindness; how, then, short of seeing a case in all its richness, can we know what its moral status is? Particularism, it seems, implies that we must be agnostic about any moral situation not immediately before us, that until we are able to see or interpret a case for ourselves, there is nothing moral we can say.

To be sure, if this were an implication of particularism, it would bode poorly for the doctrine. Such a picture is radically at odds with our picture of moral epistemic life. We commonly regard as justified all sorts of moral inferences based on discrete and limited bits of information, such as inferring that a wrong was committed when we hear a report of a robbery. Even more important, we carry a wide range of presumptions we would think it irresponsible to leave behind. It is simply not true that I greet each new situation as a moral virgin: I leave the house each morning armed with all sorts of moral presumptions, including the presumption not to go around stabbing people or operating on them without their consent.

But it is a deep misunderstanding to believe that particularism implies that discernment is the only path to moral knowledge (a misunderstanding, it must be said, that is fueled by some particularists' rhetoric).[18] Against a background of rich experience and robust patterns linking the infliction of pain with cruelty, we are often justified in projecting that pattern forward inductively. Nothing about particularism stands in the way of this view; to think that it does betrays loyalty to a model of epistemology that Wittgenstein was at pains to dislodge. Let me explain.

The idea that particularism disallows inductive claims travels by way of an assumption that the patterns we experience can be projected forward only if those patterns are backed by codifiable laws. But Wittgenstein argued that this is a mistaken model to use: very little (if anything) in epistemic life operates that way. To use the most mundane of instances: imagine that a friend needs to get a message to my neighbor Joe. Given that I have passed Joe on his front porch, where he reads the evening paper, virtually every weekday as I walk home from the subway station for the past three years, we would all regard it as perfectly appropriate for me to offer to take him the message on the grounds that I know I will see him that night. Yet any number of circumstances could prevent him from being on his porch that evening—from the newspaper not being delivered to earthquakes rocking the neighborhood to Martians beaming him aboard a spaceship—and there is no codifiable law governing the interactions among them.

It is completely misguided to think that I compute some probability function for Joe being on the porch by sifting through all the possible circumstances that could affect his presence there, weighting them by naturalistically specified probability functions, and toting up an end result.

Rather, I make judgments about whether previous patterns are robust and whether the world, while infinitely different from yesterday, is sufficiently similar in relevant ways to allow me to project that pattern forward. I come to my perfectly justified conclusion not by applying deductive laws, but by exercising competency with a variety of epistemic concepts, such as robustness, similarity, relevance, that (to give the now-familiar refrain) cannot be cashed out algorithmically. It is a *skill* to read the world, to know how to navigate through patterns of competing influences, to determine which possibilities demand investigation and which do not.

Properly understood, then, the skills crucial to the particularists' view of moral knowledge do not stop at the abilities involved in discernment or interpretation; they extend to the skillful deployment of rules of thumb. These, crucially, include knowing how to measure their limitations. It is a skill to know when you have entered a context in which previous experience no longer points the way, to know, that is, when you are on epistemically foreign ground, like a doctor whose broad experience with symptomatology is suddenly useless when she moves to a country with altogether different diseases. Or again, it is a skill to know not just how to draw the right inference from bits of information, but when you know enough to draw an inference and when you do not. Against certain epistemic backdrops, a doctor can justifiably offer a probable diagnosis based on few bits of information; in other cases, it would be negligent to do so. Against some contexts, that is, even a brief sketch of a situation assumes a shape we can read, while in others it has a gaping hole that requires one to remain agnostic, and it is a skill to know the difference.

But, again, to say that rules of thumb have limits we should stay on guard for is not to say they always meet up against those limits. To say that the moral landscape cannot be *codified* is not to say that it is *chaotic*. I can imagine a world in which we cannot presume not to stab the people one encounters (imagine your favorite post-apocalyptic movie). But that does not mean it is *our* world. Logically possible skeptical scenarios do not themselves threaten the reasonableness of our conclusions, for they may not be epistemically relevant scenarios to us, given judgments, again, about the robustness of previous patterns and the relative stability of our world in relevant respects. The next case could be precisely the one in which the infliction will not be cruel, but that does not mean I am unjustified in presuming it will not be. Against the right backdrop, then, it is perfectly appropriate to carry various epistemic presumptions and defaults. As long as the world is not morally cacophonous, we can issue moral judgments that travel well beyond individual cases.

Understanding this epistemological point should help make clear that bioethicists can be particularists and still issue policy recommendations or recommend adopting codes of conduct. It is perfectly consistent to say, in one breath, that the moral status of physician-assisted suicide depends on the details of each case, and to issue policy recommendations in the next. For we may be able to reach some inductive confidence about the sort of pattern that would emerge should the practice be made legal—say, we have good evidence that physicians would overwhelmingly be adequately careful, or, instead, that they would have too little time to assess cases to our moral satisfaction. And again, it is completely consistent to say, on the one hand, that the best way to reach moral conclusions about what to do is to deploy a highly developed moral sensibility—to eschew regulations and judge for oneself, and to urge, on the other hand, that we adopt legally binding codes of action for some parties, such as doctors. If it is predictable that too few of those parties will have the highly developed sensibility, or again, that too few will have the opportunity to see the rich details of the cases, it can be far better overall to make them adhere to a rule of thumb than to turn their impoverished or uninformed moral sense loose on the world.

Oliver Sacks tells a marvelous story about a man who suffered a curious sort of perceptual aphasia.[19] He lost the ability to recognize everyday objects like a glove, a hat, or people's faces. Devoid of the ability to see directly the objects as such, he was reduced to trying to infer what they were from the discrete bits of information he had—premises about the crude shape he could see them to have, or inductive moves he could make by reflecting on where he was or whom he was with. Since he was very smart, he arrived at the right answer surprisingly often; but clearly, his reliance on inference and presumption was a poor substitute for the ability to perceive for himself what the objects were. We tend to suffer a sort of moral aphasia. Lacking the training necessary for developing the skills needed to discern the moral landscape reliably and accurately, we end up relying on crude forms of inference. The reliance is so widespread, in fact, that theories start valorizing the deployment of inference as the height of moral maturity rather than a (useful) crutch to the blind. Such a tradition, moreover, is dangerously self-reinforcing, for when the importance of relying on principles is emphasized, skills of discernment, interpretation, and judgment can atrophy. The exemplar of moral wisdom, in fact, is one who can see or interpret for himself or herself in the elements. Failing that, we should and can rely on inference based on our skills of judging context and continuity. But we should not mistake those maneuvers for anything but the compensatory moves they are.

Notes

1 And John McDowell's wonderful discussion of Cavell. See Stanley Cavell, *Must We Mean What We Say? A Book of Essays* (New York: Charles Scribner's Sons, 1969), p. 52. John McDowell, "Virtue and Reason," *Monist* (1979): 331–50, especially sec. 4; John McDowell, "Non-Cognitivism and Rule-Following," in *Wittgenstein: To Follow a Rule*, ed. Steven H. Holtzman and Christopher M. Leich (Boston: Routledge and Kegan Paul, 1981), esp. sec. 3. The extent to which this essay is indebted to McDowell's insights will be obvious to readers of his work.

2 For Wittgenstein's own take on morality, see "A Lecture on Ethics," *Philosophical Review* 74 (1965): 3–12; and *Lectures and Conversations on Aesthetics, Psychology, and Religious Belief*, comp. Yorick Smythies, Rush Rhees, and James Taylor, ed. Cyril Barrett (Oxford: Blackwell, 1966).

3 One can, of course, be a particularist with respect to one of these levels and not the other. It is more common to find theorists arguing that relations amongst moral considerations, so identified, cannot be codified. The classic here is W. D. Ross, *The Right and the Good* (New York: Oxford University Press, 1930), and the view is explicitly advanced in Tom L. Beauchamp and James F. Childress, *Principles of Biomedical Ethics*, 4th ed. (New York: Oxford University Press, 1994).

More thoroughgoing particularists also deny that we can recover conditionals indicating the moral valence carried by any given natural property. According to this position, we cannot isolate even the prima facie moral status (such as kind or cruel) that a given nonevaluative property grounds. Representatives of this sort of particularism include Jonathan Dancy, *Moral Reasons* (Oxford: Blackwell, 1993); David McNaughton, *Moral Vision: An Introduction to Ethics* (Oxford: Basil Blackwell, 1988); McDowell, "Virtue and Reason"; and Martha Nussbaum, "The Discernment of Perception: An Aristotelian Conception of Private and Public Rationality," in *Proceedings of the Boston Area Colloquium in Ancient Philosophy*, vol. 1, ed. John Cleary (New York: University Press of America, 1986).

4 To my knowledge, Simon Blackburn is the first to use this term in this context. See Blackburn, "Reply: Rule-Following and Moral Realism," in *Wittgenstein: To Follow a*

Rule, ed. Steven H. Holtzman and Christopher M. Leich (Boston: Routledge and Kegan Paul, 1981).

5 The latter example is from McNaughton, *Moral Vision*.

6 Dancy, *Moral Reasons*, p. 61. Dancy attributes the example to Roy Hattersley.

7 Jonathan Dancy has done the most to develop this view. See his excellent discussion in *Moral Reasons*, chaps. 4–7.

8 David McNaughton, personal communication.

9 See also Dancy, *Moral Reasons*, pp. 82–83.

10 For a clear discussion, see McNaughton, *Moral Vision*, pp. 60–62, 192–94.

11 J. L. Mackie, *Ethics: Inventing Right and Wrong* (New York: Penguin, 1977).

12 Helpful discussions of this point can be found in Robert J. Fogelin, *Wittgenstein*, 2d ed. (London and New York: Routledge and Kegan Paul, 1987), chap. 11; and McDowell, "Non-Cognitivism and Rule-Following."

13 See, e.g., Wittgenstein's discussions in PI 1968, § 138–55.

14 Wittgenstein, PI, 1968, § 138–155.

15 Wittgenstein, OC 1972.

16 For representatives of theorists who speak of moral discernment, see Mark de Bretton Platts, "Moral Readings," in *Ways of Meaning: An Introduction to Philosophy of Language* (London: Routledge and Keagan Paul, 1979); McDowell, "Virtue and Reason"; Dancy, "Moral Reasons"; McNaughton, "Moral Vision"; Nussbaum, "The Discernment of Perception"; and Warren S. Quinn, "Truth and Explanation in Ethics," *Ethics* 96 (Apr. 1986): 524–44. See especially Richard Warner, "Ethical Realism," *Ethics* (July 1983): 653–79, who applies a Wittgensteinian approach to moral epistemology.

17 See Platts, "Moral Readings," and Quinn, "Truth and Exploration in Ethics."

18 See McNaughton, *Moral Vision*, and Dancy, *Moral Reasons*.

19 Oliver Sacks, *The Man Who Mistook His Wife for a Hat* (New York: Summit Books, 1985).

11   Wittgenstein: Personality, Philosophy, Ethics
Knut Erik Tranöy

"It (philosophy) leaves everything as it is"
[except philosophy and philosophers].[1]

It is no accident that "personality" is mentioned first in the title. I did have a personal contact with Wittgenstein during the last year and a half of his life. That encounter left me with strong impressions of a very personal sort.[2]

At that time I could not know that these impressions were, much later, to influence my understanding of Wittgenstein's philosophy, and his view of ethics in particular. But they did. In a literal sense, his personality impressed me before his philosophy did.

I made his acquaintance at a point in my intellectual development where Wittgenstein's philosophical ideas could not possibly make an impression on me. I was still a newcomer to philosophy. I was in Cambridge to do moral philosophy, not logic or philosophy of language. In Cambridge at that time, the prevailing and "authorized" understanding of Wittgenstein's philosophy was to read the *Tractatus* as a treatise on logic that Wittgenstein himself, in his lectures before he retired in 1947, had radically revised to become a new philosophy of language.

This was in 1949, now more than fifty years ago. Professor von Wright had only been in Cambridge for a year or so as Wittgenstein's successor. Since it was so long ago, I also feel free to relate an episode dating from before I met Wittgenstein in person. As a research student in my first term, I was shopping around for lecture courses to follow. There was in Cambridge at that time a lecturer by the name of John Wisdom (the John Wisdom of "Other Minds.") He was famous as a lecturer for the show he put on when he lectured; his classes were sometimes called "Wisdom's moaning sessions." I did go to Wisdom's lectures and found them truly remarkable. Von Wright (in whose house I was then living) was curious, but curiosity was not an academically acceptable reason for a professor to attend a colleague's classes. When I produced an imitation of a Wisdom class, von Wright's spontaneous reaction was, "But this is Wittgenstein!" When a little later I did meet the real Wittgenstein, I understood both von Wright's reaction and Wittgenstein's charisma, as well as the difference between the original and a good copy.

Looking back, I am struck by the fact that although I was completely ignorant of Wittgenstein's philosophy, he nevertheless made an indelible impression on me. In spite of the fact that Wittgenstein's philosophy did not then mean anything at all to me, that impression later turned out to be philosophically relevant, and the relevance seems to be above all of an ethical nature. This is why "ethics" comes last in the title of this essay.

The philosophical relevance of a "purely" personal impression may still be worth a moment of reflection. Many others have since pointed to the close connection between philosophy and personality in Wittgenstein, who has often been called one of the most, if not the most, important philosophers of the twentieth century. Many who have said this are outstanding professional philosophers. But to call *him* a professional philosopher sounds to me out of place. The distinction involved here is important, I think, to attempts to understand Wittgenstein as a philosopher. It is essential, I should say, to an attempt to consider him from the point of view of moral philosophy.

In any case, my impression of Wittgenstein as a person became the point of departure for my interest in his philosophy. Let me say at once, however, that I am in no way an expert on Wittgenstein's philosophy. My reading of him has been selectively guided by my interest in ethics.

It was not until around 1965, some fifteen years after his death, that I began to read Wittgenstein. In the meantime, *Philosophical Investigations* had appeared. Although the *Investigations* and other publications from his *Nachlass* were then available, in print and in translation, it was the *Tractatus* I began to read, along with the *Notebooks, 1914–16*, which had appeared in 1961. From an ethical point of view, the *Notebooks* contain more than the *Tractatus*. When the *Investigations* first appeared, many took that work primarily to be Wittgenstein's recantation of his own earlier views in the *Tractatus*. Few seemed to be interested in the *Tractatus* and its place in

Wittgenstein's thought. For my own part, I can say that I still read it with interest and profit, and not only for what it contains about ethics, religion, and metaphysics. Reading the *Tractatus* is necessary for a comprehensive and coherent view of Wittgenstein's philosophy.

I first presented my ideas about Wittgenstein as a moral philosopher in 1966 in lectures in the Universities of Durham and Newcastle. In 1973 I published that lecture as a paper, "Ethics as a Condition of the World: A Topic from Wittgenstein." [3] The main idea of that paper was to make sense of the statement that ethics is not *in* the world but is a *condition* of the world, like logic. [4] At that time, others were also beginning to voice ideas about the continuity between the early and later writings of Wittgenstein and their ethical aspects. One of the first, as far as I know, was Stephen Toulmin in his 1969 article, "Ludwig Wittgenstein." [5]

When I began to read the *Tractatus*, I was struck by what the author says about its aim and purpose. In proposition 6.54, following what he relates about throwing away the ladder after having climbed up on it, he says about the proper understanding of the Tractatus: "He [the reader] must surmount these propositions; then he sees the world rightly" ("Er muss diese Sätze überwinden, dann sieht er die Welt richtig." Ogden's translation of "überwinden" as "surmount" is not very good; "conquer" or "vanquish" might have been better.)

It is not clear what it means to see the world rightly. Two comments may be relevant, regardless of what "seeing the world rightly" means. In the first place, Wittgenstein holds seeing the world rightly to be of great importance, since it must be in some way connected with knowledge of true propositions about the world. And secondly, understanding the *Tractatus* rightly, Wittgenstein suggests, will make the reader (who might be another philosopher) see the world rightly.

The picture I had of Wittgenstein's personality from our conversations made it difficult for me to believe that he did not really mean what he wrote in the *Tractatus* about seeing the world rightly.

Of course, he may have changed his mind later about the power of philosophy to help us see the world rightly. But I cannot see that he did. He may well have thought, throughout his life as a philosopher, that philosophy *can* help us see the world rightly, in some sense. Seeing the world rightly may refer not only or not even primarily to knowledge of true propositions, but rather to an insight that in some sense (which may be the common sense) presupposes such knowledge without being, in any obvious way, deducible from it. Knowledge does not illuminate us like a friendly lightning the way insight typically does. And like knowledge, insight can also be true: it is the truth of it that is enlightening. For Wittgenstein, in the *Tractatus* and afterward, philosophy promotes insight rather than knowledge. Few words occur more conspicuously in the *Tractatus* and the *Investigations* than "clarity," "clear," and various forms of the verb "to see."

If we try to understand what Wittgenstein intended to say in the *Tractatus* and the *Notebooks*, we should also take seriously what he says about ethics. In what I have been calling the "authorized" understanding of the *Tractatus* in Cambridge around 1950, little or no attention was paid to those passages in which he speaks about ethical and religious matters. And they are in fact quite a few. To overlook or refuse to consider them and thus also the topic of seeing the world rightly as a problem for philosophy seemed to me a strange thing in light of the admiration Cambridge philosophers had for Wittgenstein.

I began to look more systematically for remarks about ethics in the *Tractatus*, the *Notebooks*, and, later, the *Investigations*. The really challenging statement, found in both the *Tractatus* and the *Notebooks*, is the idea that not only logic but ethics, too, are conditions of the world. The clearest expression of this is the one I quoted above: "Ethics does not treat of the world. Ethics must be a condition of the world, like logic" (NB 24.7.16.). But there is no coherent argument about ethics as one may surely say there is about logic.

Toward the end of the *Notebooks* he does, in fact, say that now finally the connection between ethics and the world is to be clarified (NB 9.10.16). He never did that. But the scattered and aphoristic remarks about ethics are nonetheless suggestive and challenging. In the first place, there are two concepts of "world" in the *Tractatus*. The opening proposition tells us that the world is everything that is the case. Logic can then, obviously, be understood as a condition of the world in that sense of "world." It seems unlikely, however, that Wittgenstein should want to say that ethics is a condition of "alles, was der Fall ist." In the *Notebooks*, however, he uses "world" in a different sense, more like Husserl's "life-world/Lebenswelt":

"Die Welt und das Leben sind eins" (The world and life are one; T sec. 5.621). And in T 5.63 he says, "Ich bin meine Welt" (I am my world). The ambiguity of the term "world" in the *Tractatus* could also, I suppose, be taken to indicate or betray a lack of consistency in the argument of the *Tractatus* as a whole. But the relationship between these two concepts of world is, indeed, problematic, not least from the point of view of ethics. (Recall "Hume's law" and the many who thought and think there can be no logical relation between "is" and "ought.")

Today, however, it may be easier than before to make sense of the idea that ethics is a condition of the world. In somewhat the same sense in which Wittgenstein said that logic is prior to any experience (T sec. 5.552), we might now say that the *conceptual repertoire* of morality is constitutive of human morality, that it is prior to any discourse about moral matters and thus also prior to all moral agreement and disagreement. What I mean by the "conceptual repertoire of morality" is the universal occurrence in human languages of terms and concepts for such things as, for instance, what a promise is, and what a normative law or rule is in contrast to a law of nature (which Wittgenstein touches on in T sec. 6.422). There would be no moral issues about which to agree and disagree if we did not somehow share such terms and notions as right and wrong, good and bad, promise and contract, rights and duties.[6] Briefly, morality as an institution is part of what is needed to make a world.

Much depends, then, on the meaning we give to the concept of "world." The idea of ethics as a condition of the world is interesting not simply because Wittgenstein proposed it, but in its own right. It may be called impressive that Wittgenstein had this insight (of ethics as a condition of the world) eighty years ago. No wonder he disliked the use the Vienna Circle made of his statement that "ethics cannot be expressed" because "there can be no ethical propositions" (T secs. 6.421, 6.42). His reasons for saying this must have been very different from the reasons that made the Vienna Circle applaud him.

The coherence in Wittgenstein's philosophy, as I see it, is partly of an ethical nature and concerns (also in part) the aim, purpose, and justification of philosophy. According to the *Tractatus*, it is important to see the world rightly. In the *Investigations*, the importance of philosophy is linked to the notion of therapy. If one sees the world wrongly and it is important not to, then one may need help to see it rightly, especially if the confusion is deep, as Wittgenstein no doubt thought it was.

There is one passage in particular in the *Investigations* that I have always found challenging and even bewildering: "The real discovery is the one that makes me capable of stopping doing philosophy when I want to.—The one that gives philosophy peace, so that it is no longer tormented by questions which bring *itself* in question. . . . There is not *a* philosophical method, though there are indeed methods, like different therapies" (PI § 133). In the same paragraph he says that he is aiming at a *complete* clarity which would make the philosophical problems *completely* disappear.

Perhaps in the 1920s, after he had published the *Tractatus*, Wittgenstein did think he could stop doing philosophy. Perhaps he did try to throw the ladder away behind himself and then discovered that the clarity he had reached was not complete after all.

The *Tractatus* ends on the note of seeing the world rightly. The notion of world is not conspicuous in the *Investigations*, but the idea of clarity is, and so is that of insight. That is the reason I added the comment to the initial quotation about philosophy leaving everything as it is. Philosophy must do something to, for, or with philosophers, and there is very little doubt that Wittgenstein has done a great deal to philosophy. For him, then, it would be logical to warn against philosophy if he thought it could do the philosopher harm, as in fact he thought and so warned his pupils. In the end, however, philosophy had done something good for him if it had made his life "wonderful."

"Therapy" comes from a Greek word that may mean both "to nurse" and "to cure." Therapy—the replacement of "seeing the world rightly"—is not only surgery and curative medication. If there is one branch of medicine from which Wittgenstein's therapy metaphor may draw support, it is psychotherapy, which is equally long lasting and of uncertain outcome. This brings to mind the oft-quoted remark from the *Tractatus*: "The solution of the problem of life is seen in the vanishing of this problem" (T sec. 6.521).

I turn once more to the link between personality and thought in Wittgenstein. The strength and depth of my impression of him as a person were increased by concomitant impressions of character traits of a moral nature. He certainly had what might be called an uncompromising attitude to truth. To say that he had a deep respect for the truth—and a confident belief in clarity as a means of making the truth visible—is to say too little. There was a passionate independence and seriousness about his thinking, even when he talked about non-philosophical matters—about Ibsen's dramas, for instance, where I (being a Norwegian) was after all in a position to understand more of what he said. And some of the things he said about philosophers were not so difficult to understand, either. He once said that Russell was intellectually more gifted than Moore, and that despite this, it was still (this was in 1950) possible to have a fruitful discussion with Moore but not with Russell because Moore had something Russell lacked: sincerity. And Russell, in fact, has said that Wittgenstein had more of that than anybody else: "No one could be more sincere than Wittgenstein or more destitute of the false politeness that interferes with truth." [7]

Perhaps it is the seriousness and sincerity with which we do the thinking that makes the difference, a sincerity children may have more often than adults. That may be one of the reasons why there are special dangers connected with being a *professional* philosopher. In today's language, in many circles, to be "professional" is high praise. But that adjective does not seem to fit Wittgenstein at all. I am willing to believe that moral seriousness and sincerity were at the core of his intellectual genius as a philosopher: an uncompromising respect for the truth that resulted in an equally uncompromising demand for intellectual decency down to the least detail. Which was also, I think, something that made him at times a demanding friend.

Personally, I strongly feel that an attitude of respect for truth is one of the fundamental features of human morality. And the capacity for morality is essential to our humanity: this is, indeed, another aspect of ethics as a condition of the world, and not only the world of academics. It is not the practical, instrumental, means-end value of true knowledge, or the sheer wrongness of lying, which I now think of as morally fundamental. What I have in mind is a non-consequentialist aspect, if I may put it like that, or, with a quotation not from Wittgenstein but from Rawls, "Being first virtues of human activities, truth and justice are uncompromising." [8]

There is also the remarkable, empirical fact that the successful search for truth and the acquisition of true knowledge can themselves generate one of the finest joys of which man is capable, the joy that Augustine called *gaudium de veritate*, joy at the possession of truth.[9] In its most perfect form it is to be attained, for Augustine, in the beatitude of the afterlife. But we also know, and Augustine knew, that it may be attainable here and now as well, and in a measure that approaches or may become ecstasy. I say this not in defense of rationalist or intellectualist forms of ethics, or as an argument against the claims of other, more "emotive" aspects of morality, but as a statement about human nature, and to emphasize the fundamental role of specifically human forms of reflective consciousness in the interaction of between cognitive and other, less cognitive human powers and activities. Joy at the possession of truth is, at the same time, the remarkable union of the cognitive and the emotive where (to use medieval language) the object of love and the object of knowledge coincide in one and the same object: the truth.

In an earlier attempt to review my image of the ethical profile of Wittgenstein's personality, I tried to apply to him the distinction between deontological and consequentialist ethics.[10] I confess that I have been, and still am, inclined to perceive him as being mainly of a deontological bent: his own duties were more important to him than his rights or the rights of others. Norman Malcolm seems to have had a similar view. He talks about "his absolute, relentless honesty" and "his ruthless integrity, which did not spare himself or anyone else."[11] In his autobiography Russell says, "He was perhaps the most perfect example I have ever known of genius as traditionally conceived, passionate, profound, intense, and dominating. He had a kind of purity which I have never known equalled except by G. E. Moore."[12]

I no longer feel sure that this distinction (between deontology and consequentialism) is helpful in an attempt to understand Wittgenstein as a moralist. His views of ethics are not systematic. In the *Tractatus*, there are remarks on laws as well as on values. Perhaps the very distinction between deontology and consequentialism is a product of the makers of moral theories, intended for use in the characterization of theories. Moral principles and theories may be either deontological or consequentialist. People, however, are not the one or the other—mostly they are both, and the mix differs from one person to the next. Some people are more inclined to stress and seek support from normative rules and principles, and from various forms of justice in particular. There are others for whom values such as love, forgiveness, tolerance, and compassion are at the heart of morality. Although Wittgenstein might be an extreme representative of the former kind, no person who is unique, independent, and complex in the extreme lends himself to smooth classification.

In my concluding remarks I turn to Wittgenstein's last words to Mrs. Bevan: "Tell them I've had a wonderful life" (first quoted, as far as I know, in Malcolm's *Memoir*.) There is an ethical aspect to this remark as well, with an extension from general ethics to bioethics. Malcolm did not at first understand how he could say such a thing when his life had been so "fiercely unhappy." It should not be difficult to make sense of Wittgenstein's last remark if we see it as his own definitive judgment on the quality of his own life—and "over a complete life," to cite a remark by Rawls on the good life.[13] Perhaps Malcolm, at that time, had too narrow a concept of happiness. But it could also be that a good life is not necessarily the same as a happy life. In either case, we might say that quality or goodness of life "over a complete life" does not uniquely depend on what we often call happiness.

In the first place, a common view of happiness—as feeling good all or most of the time, which is now perhaps the dominant view in our culture—was hardly a view shared by Wittgenstein. And even on the view that, essentially, happiness is a balance of pleasure over pain, there is such a thing as the final settling of accounts in which good states and events occurring late in life may count for more than bad ones earlier. The importance of the quality of the last stretch of conscious life is almost universally acknowledged in our culture. Proof of this may be found in the death announcements in (Nordic) newspapers. There is truth in the old German (and Nordic) saying *Ende gut, alles gut* (when the ending is good, all is good), even if it is not the whole truth. Wittgenstein's unique authority in this matter is unquestionable. He was the only one who had lived Ludwig Wittgenstein's life. Of course, the point applies to each and every one of us. No matter how we define happiness or a good life, it is only the person who has lived the life in question who has the competence and the moral authority to make this final judgment. To deny this might seem to entail a strange kind of paternalism. Others cannot even correct the dying person's own definitive judgment, provided the person was of sound mind when he made it.

Wittgenstein's last words also provoke other reflections. We know that he was present when his father died and that in a letter to Russell he wrote that his father's death was "the most beautiful death I can imagine . . . this death was worth a whole life." And McGuinness adds, "A good life should issue in a death one could welcome: perhaps, even, a good life *was* one that led to a good death." [14]

These remarks contain suggestions of two rather different conceptions of what a good life is. One conception focuses primarily on pain and pleasure, misery and happiness, and nobody would deny that these are important and morally relevant features of human lives. The other focuses on something else that is not easy to express in words—perhaps above all on meaning and meaninglessness, on achievement and defeat. But it also takes into account such a thing as the Augustinian *gaudium de veritate*, which may well be called a particular form of happiness, and a happiness, furthermore, that is not for sale. The difference between these two perspectives is not only philosophically interesting. It is practically important as well. The first, happiness-related alternative is something it is possible to aim at and to plan for—barring the *gaudium de veritate*—and where others may be able to help: care and pain control at the hands of doctors and nurses, for instance. Controlling and promoting the quality of life is an important dimension of contemporary medicine, most clearly manifested in the hospice movement. Considerable efforts are now made in medicine to control and enhance quality of life in this sense. Palliative medicine, the product of special knowledge and skills, is used in a goal-directed way for the benefit of helplessly suffering patients. The purpose of palliative medicine is, briefly and perhaps inadequately, to improve quality of life through control of pain and the many other forms of physical and mental suffering. It is interesting and not morally irrelevant to note that the use of certain kinds of palliative drug medication in other situations may be equivalent to substance abuse.

The other kind of good life is tied, I suggest, to the dimension of meaning and meaninglessness. A good life in this sense—and over a complete life—is not something others can plan, organize, and control at all. It is even doubtful in what sense the one whose life it is can do that, although, as a rule, he or she has to take responsibility for it. As Wittgenstein did. (This holds for normal adults, but not necessarily for children.) If one should experience moments of *gaudium de veritate*, and recurring moments, they often come with the spontaneity and force of an apparently irresistible insight, although it may be the reward of years of effort. It is not something one can plan for as a foreseen and distant goal. It is beyond the reach of medical good and evil. Perhaps it is only available as a welcome but unintended side effect of other aims, efforts, and achievements. It is certainly not the quality of life palliative medicine and medical ethics can control and deliver. The satisfaction and misery a search for this kind of quality may generate cannot be observed and appreciated by outsiders, in contrast to the sufferings and satisfactions connected with the active search to maximize happiness-related welfare.

Is it amiss to think that a happiness-related quality of life is closer, perhaps, to the consequentialist family of ethics, while the meaning-related quality of life is more closely related to the deontological family? In his persistent search for clarity and insight, Wittgenstein was not, it seems, primarily concerned with the consequences, for himself or for others, of his philosophical efforts to reach "complete clarity." His letters and diaries seem to be persuasive evidence that it was intellectual or spiritual achievement and frustration that, more than anything else, were the determinants of happiness and misery in his life. And Malcolm, after he changed his mind about Wittgenstein's last words, added, "I find it impossible to believe that this activity of creation and discovery gave him no delight, even though he always felt that it came short of what was needed." [15]

That is what I have in mind when borrowing the Augustinian confession about joy at the possession of truth.

Notes

1 The epigraph is from Wittgenstein, PI § 133; the comment is mine.
2 Cf. K. E. Tranöy, "Wittgenstein in Cambridge, 1949–51: Some Personal Recollections," *Acta Philosophica Fennica: Essays on Wittgenstein in Honour of G. H. von Wright*, 28, nos. 1–3 (1976): 11–21.
3 "Ethics as a Condition of the World" was first printed in *Norsk filosofisk tidsskrift* in 1973. It was reprinted in Andrew J. I. Jones, ed., *The Moral Import of Science* (Bergen: Sigma forlag, 1988), 157–70.
4 "Die Ethik handelt nicht von der Welt. Die Ethik muss eine Bedingung der Welt sein, wie die Logik." NB 24.7.16.
5 Stephen E. Toulmin, "Ludwig Wittgenstein," *Encounter* (Jan. 1969): 58–71.
6 I realize, of course, that ancient Greek words are not perfectly synonymous with those English (or Norwegian) words we use to translate them. But their meanings

"overlap" with contemporary meanings to such an extent that we can, after all, communicate about shared problems and the misunderstandings that translations give rise to. There is, of course, a similar problem when we translate from one modern language to another. Martha C. Nussbaum makes interesting observations about the translation of Greek *eudaimonia* into English in *The Fragility of Goodness: Luck and Ethics in Greek Tragedy and Philosophy* (Cambridge: Cambridge University Press, 1986), p. 6.

7 The quotation is from Brian McGuinness, *Wittgenstein: A Life* (London: Duckworth, 1988), where Russell is quoted on p. 102.

8 John Rawls, *A Theory of Justice* (Harvard University Press, 1971), p. 4.

9 K. E. Tranöy, "Aquinas," in *A Critical History of Western Philosophy*, ed. D. J. O'Connor (New York: Free Press of Glencoe/Macmillan, 1964), pp. 98–123.

10 K. E. Tranöy, "Wittgenstein: Ethics and the 'Wonderful Life,'" in *Wittgenstein: Towards a Re-Evaluation*. Proceedings of the 14th International Wittgenstein-Symposium, ed. R. Haller and J. Brandl (Vienna: Verlag Hölder-Pichler-Tempsky, 1990), pp. 273–79.

11 Norman Malcolm, *Ludwig Wittgenstein: A Memoir* (Oxford: Oxford University Press, 1984), p. 26.

12 Here quoted from McGuinness, *Wittgenstein: A Life*, p. 100.

13 John Rawls, *Political Liberalism* (New York: Columbia University Press, 1993), p. 6.

14 McGuinness, *Wittgenstein: A Life*, p. 166.

15 Malcolm, *Ludwig Wittgenstein: A Memoir*, p. 84.

Notes on Contributors

Larry R. Churchill is Professor of Social Medicine at the University of North Carolina at Chapel Hill and codirector of the newly established Center for Health Ethics and Policy. His major intellectual and research interests concern social justice in health care, the ethics of research with human subjects, and the moral dynamics of care at the end of life.

David DeGrazia is Associate Professor of Philosophy at George Washington University, where he has taught since 1989. He is the author of *Taking Animals Seriously: Mental Life and Moral Status* (Cambridge University Press, 1996) and coeditor, with Thomas A. Mappes, of *Biomedical Ethics*, 5th ed. (McGraw-Hill, 2001). Currently his major research interests are at the interface of personal identity theory and bioethics.

Cora Diamond is Kenan Professor of Philosophy and University Professor at the University of Virginia, and has also taught at the University of Aberdeen and Princeton University. She is the editor of *Wittgenstein's Lectures on the Foundations of Mathematics, Cambridge, 1939* (Cornell University Press, 1976). Her recent work includes *The Realistic Spirit: Wittgenstein, Philosophy, and the Mind* (MIT Press, 1991) and articles on Frege and Wittgenstein, philosophy of language, ethics, and philosophy in relation to literature.

James C. Edwards is Professor of Philosophy at Furman University, where he has taught since 1970. He is the author of *Ethics without Philosophy: Wittgenstein and the Moral Life* (University Press of Florida, 1982) and *The Plain Sense of Things: The Fate of Religion in an Age of Normal Nihilism* (Pennsylvania State University Press, 1997).

Carl Elliott is Associate Professor of Pediatrics and Philosophy at the University of Minnesota. He is the author of *A Philosophical Disease: Bioethics, Culture, and Identity* (Routledge, 1998) and coeditor, with John Lantos, of *The Last Physician: Walker Percy and the Moral Life of Medicine* (Duke University Press, 1999).

Grant Gillett is Professor of Medical Ethics at the University of Otago and a practicing neurosurgeon. He gained a D.Phil. in philosophy at Oxford University and was appointed to a fellowship at Magdalen College in 1985. From there he moved to the University of Otago and has continued to teach and write in philosophy, philosophical psychology, and philosophy of psychiatry. His latest book is *The Mind and Its Discontents* (Oxford University Press, 1999). He has also edited several books and published widely in the areas of philosophy of mind and bioethics.

Paul Johnston has a D.Phil. in philosophy from Oxford University and is the author of *Wittgenstein and Moral Philosophy* (Routledge, 1989), *Wittgenstein: Re-Thinking the Inner* (Routledge, 1993), and *The Contradictions of Modern Moral Philosophy* (Routledge, 1999).

Margaret Little is Senior Research Scholar at the Kennedy Institute of Ethics and Associate Professor of Philosophy at Georgetown University. She is coeditor, with Brad Hooker, of a collection of new essays on particularism and generality in ethics entitled *Moral Particularism* (Oxford University Press, 2001). She is currently finishing a book entitled *Abortion, Intimacy, and the Duty to Gestate* (Oxford University Press, forthcoming).

James Lindemann Nelson is Professor of Philosophy at Michigan State University and also teaches in MSU's medical schools through the university's Center for Ethics and Humanities in the Life Sciences. He is the coauthor of *The Patient in the Family* and coeditor of *Meaning and Medicine: A Reader in the Philosophy of Medicine*. His essays with a special focus on Wittgenstein have appeared in the *Proceedings of the Austrian Ludwig Wittgenstein Society* and in *Dialogue: Canadian Philosophical Review*.

Knut Erik Tranöy is Emeritus Professor of Philosophy and Medical Ethics at the University of Oslo. He is currently completing a book in English on the importance of common morality for a more realistic understanding of the moral foundations of medical ethics.

Index

Library of Congress Cataloging-in-Publication Data
Slow cures and bad philosophers: essays on Wittgenstein,
medicine, and bioethics / edited by Carl Elliott.
Includes bibliographical references and index.
ISBN 0-8223-2657-4 (cloth : alk. paper) — ISBN 0-8223-2646-9 (pbk. : alk. paper)
1. Medical ethics—Philosophy. 2. Wittgenstein, Ludwig, 1889–1951.
3. Bioethics—Philosophy. I. Elliott, Carl, 1961–
R725.5 .S58 2001   174′.2′01—dc21   00-063661